Matthew Klemm

**Medical Anthropology in the Late Middle Ages**

# Medicine in the Medieval Mediterranean

Edited by
Alain Touwaide



## Volume 11

Matthew Klemm

# Medical Anthropology in the Late Middle Ages

Pietro d'Abano on Body, Soul, and the Virtues

DE GRUYTER

ISBN 978-3-11-061134-2
e-ISBN (PDF) 978-3-11-061263-9
e-ISBN (EPUB) 978-3-11-061152-6
ISSN 2569-314X

**Library of Congress Control Number: 2024945240**

**Bibliographic information published by the Deutsche Nationalbibliothek**
The Deutsche Nationalbibliothek lists this publication in the Deutsche Nationalbibliografie; detailed bibliographic data are available on the internet at http://dnb.dnb.de.


Cover illustration: Redrawings of two enamels on the Holy Crown of Ungary, Byzantium, ca. 1070 A.D.
Typesetting: Integra Software Services Pvt. Ltd.
Printing and binding: CPI books GmbH, Leck

www.degruyter.com
Questions about General Product Safety Regulation:
productsafety@degruyterbrill.com

FSC www.fsc.org MIX Papier | Fördert gute Waldnutzung FSC® C083411

To Sarah

**Illustration 1:** The unbalanced figure of Inconstancy from the Scrovegni Chapel, painted by Giotto along with the other vices 1303–1305 in Padua, where Pietro was also completing his major works. Mondadori Portfolio/Archivio Antonio Quattrone/Bridgeman Images.

# Acknowledgements

This book began long ago as my PhD dissertation (2007) at Johns Hopkins University, and then sat for a number of years while I was distracted by other things, so my debts go back many years. Without the encouragement and careful reading of Alain Touwaide, the editor of the *Medicine in the Medieval Mediterranean* series, it may never have been completed.

Several entities and grants supported this research, including the William and Lois Diamond Fellowship and a Charles S. Singleton award at Johns Hopkins, which funded a research trip to the Vatican. Grants from the Gladys Krieble Delmas Foundation made research in Padua possible. The Belgian American Educational Foundation and the De Wulf-Mansion Center provided support for my stay in Leuven. The Friends of the University of Wisconsin Libraries funded summer research in Madison. Summer research grants from Ithaca College have supported my work more recently. I thank Brill Publishers for permission to absorb portions of my previous work into this one.

There are many people to whom I owe thanks for their help and encouragement. These include two readers when this book was first reviewed at Ashgate, who provided many helpful comments. I am thankful to Davide Cellemare and Aurlién Robert, who invited me to present research which later became part of this book. Among others, I especially thank Craig Martin and Bill Duba for their friendship and suggestions over the years.

Katherine Tachau first inspired me to study the intellectual history of the Middle Ages when I was an undergraduate at the University of Iowa, and continued to supply encouragement and criticism long after I left, eventually serving on my dissertation committee at Johns Hopkins. Without her mentorship, none of this would have been possible. At Johns Hopkins, I am grateful to Lawrence Principe, Gabrielle Spiegel, and especially Nancy Struever, who first pointed me in the direction of Pietro d'Abano and gave my thesis a shape when my thoughts were very scattered.

Much of my early work on Pietro d'Abano began at the De Wulf-Mansion Center at KU Leuven, thanks to Maarten Hoenen, who brought me there as part of a project on Pietro. Russell Friedman provided a friendly environment during my time there and is always welcoming when I return. In Leuven, my colleagues Pieter Buellens, Gijs Coucke, Guy Guldentops, and Tine Swaenepoel, and others made my stay enjoyable and productive. I am especially grateful to Pieter De Leemans, who shared my interest in medieval animals, and is greatly missed.

More recently, I am grateful to my colleagues at the Ithaca College History Department past and present, including Jonathan Ablard, Penny Bogardus, Karin Breuer, Vivian Bruce Conger, Jason Freitag, Zoe Shan Lin, Pearl Ponce, Michael Smith, Michael Trotti, and Zenon Wasyliw, along with Alexis Becker, Dan Breen, and Bob Sullivan in English, who have all endured hearing about Pietro d'Abano and other aspects of the middle ages and have aided my work in various ways. They continually make my job at Ithaca College enjoyable.

https://doi.org/10.1515/9783110612639-202

My father David, and my mother, Joyce, inspired my interest in the humanities in their very different ways. I am grateful to my sister, Hannah, and especially my brother, Stephen Klemm, whose comments on several chapters were extremely helpful. My greatest debt is to my family for their patience while this work was completed, including my children, Frances and Oliver, and my wife Sarah von Schrader, to whom this book is dedicated.

# Contents

**Acknowledgements — IX**

**List of Illustrations and Table — XIII**

**Introduction — 1**
Pietro's Life and Thought — **2**
Pietro's Works — **9**

**Chapter 1**
**Body-Soul Relations in Medical Philosophy from Salerno to Pietro d'Abano — 17**
The Long Twelfth Century — **18**
New Translations and Commentaries — **18**
Philosophical Uses of Medicine — **25**
Critics — **28**
Scholastic Medicine in the Early Thirteenth Century — **30**
Changing Attitudes — **30**
The First Decades of Medicine in the Universities — **31**
Aristotelian Natural Philosophy — **34**
New Ambitions, 1270–1320 — **37**
Medicine as Science — **38**
Galenism and Aristotelianism — **40**
Philosophy at Paris — **41**

**Chapter 2**
**Complexion between Body and Soul — 45**
Aristotle and Galen on Complexion — **47**
Pietro's Reconciliation — **50**
Complexion and Substantial Form — **53**
The Best Complexion — **57**

**Chapter 3**
***Spiritus*, Life, and the Generation of the Soul — 61**
*Spiritus* — **61**
Life and the Generation of the Soul — **67**

**Chapter 4**
**The Organic Soul and the Intellect — 75**
The Organic Soul — **75**
The Intellect and its Parts — **83**

**Chapter 5**
**The Power of Things: Confidence and Incantations — 93**
Confidence as Cure — **94**
Incantations and the Power of Words — **100**

**Chapter 6**
**The Physiology of Moral Virtue — 111**
Aristotle and the Philosophical Import of the *Problemata* — **111**
*Particula* 27: Irascible Virtue — **115**
*Particula* 28: Concupiscible Virtue — **119**
*Particula* 29: Justice — **123**

**Epilogue — 131**

**Bibliography — 133**

**General Index — 151**

# List of Illustrations and Table


**Illustration 1** Giotto, Inconstancy, Scrovegni Chapel, Padua. Mondadori Portfolio/Archivio Antonio Quattrone/Bridgeman Images —— **VII**

**Illustration 2** Ring of Complexion, *Conciliator*, f. 30r (Venice: Luceantonii de Giunta, 1520). Courtesy of the Countway Library of Medicine Rare Book Room at Harvard Medical School. —— **51**

**Table 1** *Conciliator,* diff. 57: *Catena de viribus animae*: Chain of the Soul's Faculties —— **78**


https://doi.org/10.1515/9783110612639-204

# Introduction

Pietro d'Abano was one of the most celebrated and notorious thinkers of the late Middle Ages, famous for his work in medicine, philosophy, and astrology. His ambitious combination of these disciplines contributed to several areas of inquiry in scholastic thought and his work remained influential well into the seventeenth century. This study concerns a central aspect of his thought which combines elements of all his interests, what I am calling his "medical anthropology". The term "anthropology", aims to capture a range of inter-related topics fundamental to Pietro's philosophy of human nature, including basic physiology, psychology, moral philosophy, and the activity of the imagination and intellect. My claim that this anthropology is distinctively "medical" reflects the fact that Pietro most thoroughly developed many of these ideas within a traditional medical framework in his *Conciliator*, which he completed in 1310, his massive "reconciliation" of medicine and philosophy. While not technically a materialist anthropology, this label also calls attention to his prioritization of the material, mutable, naturalistic, or contingent aspects of human life – the domain of the *medicus*.

In his work, Pietro develops fundamental principles of medical theory to explain the integration of physical and spiritual capacities in humans, and of physiological and mental processes. Thus, his discussion of the medical notion of "complexion", the mixture of physiological components – blood, phlegm, yellow and black bile – and their qualities in the body which provides the explanatory basis for health and sickness, becomes an account of internal process and change that explains excellence and dysfunction in all human activities, including the higher operations of the soul. In this, his anthropology challenged the dominant concerns of Christian anthropologies, which, in his opinion, failed to recognize the full implications of the hylomorphic union of the soul to the material body. I argue that this anthropological initiative should not be regarded as a digression from his role as physician; rather, it is central to his entire intellectual project. As Pietro remarks early in the *Conciliator*, "even if a physician does not consider the soul directly, it is, nevertheless, his true object."[1]

From the perspective of the critical tradition transmitted most famously by Francesco Petrarch (1304–1374) in his *Invective contra medicum (Invectives against a Doctor*) (1355), Pietro's philosophical interests might seem to epitomize the worst features of medieval scholastic medicine: that it was abstract, wordy, and ultimately unable to pro-

1 *Conciliator*, diff. 5 (f. 9v, col. 1, F).: "Et si medicus animam non consideret directe, verum eius amplius subiectum." Aristotle (384–322 BCE) frequently makes similar comments in his biological works, saying that when studying the material structure of the body, the natural philosopher must also take the soul into account. See Aristotle, *De partibus animalium* I.1 = 641a29–33 (ed. and Engl. transl. Peck 1937: 71). On the soul as an object of investigation in natural philosophy, see Zupko 1997, and De Boer 2013.

https://doi.org/10.1515/9783110612639-001

duce real cure.[2] Yet if we set look past this polemic, they illustrate the range of possibility in medical thought during what is now recognized to have been a highly creative and formative period in the history of medicine.[3] Indeed Petrarch's hostility highlights the fact that medicine had emerged as a prestigious, professional, and lucrative occupation in the two generations prior to the publication of his *Invective*. We can point to concrete accomplishments such as the establishment of formal medical education at the schools, the increase of licensed practitioners, the sanctioning of human dissection, an acceleration in the publication of new medical works, or the translation and diffusion of great numbers of Greek and Arabic sources essential for defining medicine as an academic discipline, many of which might seem more in line with modern notions of the progress of medicine.[4] Yet in the medieval context, these developments were highly dependent on the increased prestige made possible by the efforts of physicians to associate medical theory with the greater intellectual standing of philosophy.[5] At the turn of the fourteenth century, this theoretical and abstract side of medicine represented a new and sustained affirmation of the status of medicine as a field of knowledge in the universities. While many other physicians of his generation participated in this movement, Pietro's philosophical commitments and his judgment of the realm of medical expertise surpassed most, if not all, of his contemporaries.[6]

## Pietro's Life and Thought

The precise details of Pietro's life are difficult to pin down, but the most significant fact is that his career spanned diverse intellectual scenes north and south of the Alps. He was most likely born in 1250 in the town of Abano, near Padua in the Veneto.[7] Little is

2 The long tradition of scorn for medieval medicine has caused more than one modern scholar to practically apologize for taking it as an object of serious study. As Jacquart asks, "In the entire history of universal culture, is there any other field that has more aptly served as a symbol of sterility of thought than that of scholastic medicine?" (Jacquart 1993/1998: 197). Francesco Petrarca, *Invective contra medicum* in *Invectives*. (ed. and Engl. transl. Marsh 2003: 2–179).

3 On the formative nature of the period spanning roughly from 1270–1320, see especially Siraisi 1981, Ottosson 1984, Agrimi and Crisciani 1988, and Jacquart 1993/1998.

4 For a summary of these and other developments at the medieval universities, see Siraisi 1992 and, on medicine in general, Siraisi 1990.

5 For a work that focuses on the professional and monetary benefits derived from this association, see French 2003.

6 Siraisi 1985: 161, stresses the unusual strength of Pietro's philosophical interests: "Yet it is obvious that during the two centuries following Pietro's death, philosophical interests at anything like Pietro's level of seriousness were characteristic of only a relatively small segment of the Italian medical community." For some of the philosophical interests of other physicians of Pietro's generation, see Siraisi 1981.

7 For more complete biographies of Pietro, see Thorndike 1923: 874–939, and Paschetto 1984: 19–34, which includes references to older biographical literature. Paschetto 1984 22–26, reviews the evidence and offers 1247–1258 as the range of possibility for Pietro's birth, suggesting 1250 as most likely.

known about his early life until he can be placed in Paris by at least in 1293. At Paris, he began his most significant works, including the *Conciliator*, the *Expositio Problematum Aristotelis*, and the *Lucidator*. This period in Paris probably accounts for some of the philosophical interests not shared as intensely by many of his contemporaries. His writings reflect some of the preoccupations of the faculties of theology and the arts there and they reveal the presence of continuing naturalistic interests at Paris in the tradition of Albertus Magnus (ca. 1200–1280) and Roger Bacon (ca. 1219–1292), both major influences on Pietro.[8] During his time in Paris, or perhaps between Paris and his return to Padua, Pietro travelled to Constantinople where he learned Greek.[9] He returned to Padua by 1302 where he had been recruited by the commune to teach medicine, philosophy, and astrology at the university.[10] Here he completed the *Conciliator*, *Expositio*, and *Lucidator*, publishing the final versions of all of these massive works in 1310. He probably died in 1315. From his will, we know he had children.[11] Most important among the other details of his life, he tells us that he was investigated at some point in Paris by a Dominican inquisitor, who had identified fifty-five suspect ideas in his work, including the materialist position that the "intellective soul is brought about from the potency of matter", which he vehemently denied, and that he was rescued by the intervention of the Pope.[12] He may have been brought to trial twice more after his return to Padua.[13] According to at least one witness, he was posthumously found guilty of heresy and his bones were disinterred and burned.[14]

---

**8** Alessio 1976.

**9** De Leemans 2007: 108–109. De Leemans revises earlier accounts that place Pietro's trip to Constantinople between 1270–1290, before he went to Paris (e.g, Paschetto 1984: 26). However, Pietro states in the prologue of his *Expositio Problematum* that he travelled to Constantinople to learn Greek after he had already begun commenting on the text. This suggests that the trip happened either during his lengthy stay in Paris or after Paris and before he returned to Padua in 1302.

**10** On Pietro's employment by Padua, see Thorndike 1923: 930–933. For his foundational role in the school at Padua, see Siraisi 1973.

**11** For an edition of Pietro's will, see Pesenti 1980.

**12** Jacquart 2016 argues convincingly that the most likely source of this particular accusation is Pietro's discussion of the so-called *intellectus vocatus*, originally in an early version his *Compilatio physionomie*. On the *intellectus vocatus* and this accusation, see Chapter 3: 72–73 and Chapter 4: 90 *Conciliator*, Diff. 48 (f. 71v, col. 2, G): "Et ideo apparet hic erroneus intellectus Jacobitarum me persequentium tamquam posuerim animam intellectivam de potentia educi materiae, cum aliis mihi 54 ascriptis erroribus. A quorum manibus gratia dei et apostolica mediante laudabiliter evasi." Pietro does not mention other errors specifically, however, in the *Conciliator*, diff. 9 (f. 15v, col. 2, F), in the context of a discussion of astrology, Pietro refers to accusations that he had been "derogating divine wisdom", again adding that he had been saved by "apostolic intervention".

**13** For the murky evidence for these procedures, see Paschetto 1984: 29–32; Thorndike 1923: 938–947; and 1926.

**14** Thomas of Strasburg (d. 1357), General of the Augustinian Order 1345–1357, reports a generation after Pietro's death in his *Sentences Commentary* that he witnessed the burning of Pietro's bones. See Thorndike 1923: 943–944.

Pietro's renown in the centuries after he wrote has not resulted in an equivalent place in intellectual history today, although there has been a renewal of interest in him in the last twenty years.[15] He is best known in the history of medicine,[16] and related topics such as pharmaceuticals and poisons,[17] astrology,[18] and magic and occult properties.[19] He has also earned attention as a translator.[20] Although he obviously regarded himself as a philosopher (referring to himself as *Petrus Paduanus, philosophie minimus alumpnorum* [sic] "least of the students of philosophy", with false modesty),[21] the nature of his philosophy does not exactly correspond to modern philosophical categories.[22] Some exceptions include his association with the origins of Averroism in Padua for his alleged role in transporting "Latin Averroism" from Paris to Padua,[23] and for his contributions to scientific methodology.[24] Otherwise, he has proven to be too eclectic and probably too committed to the occult to be taken seriously by most modern historians of philosophy.[25] Because of the breadth of the authorities cited by Pietro, he can be used as a guide to the various naturalistic texts available in his time.[26] Never one to be shy about his own merits, he often calls attention to the scope of his learning, as in the

---

15 Since Paschetto 1984, see the excellent and entertaining study by Joan Cadden (Cadden 2013) and three collections of essays devoted to Pietro: a special issue of *Medicina nei secoli* 2008, Boudet, et al. 2013, and De Leemans and Hoenen, 2016. See especially, Boudet et al. 2013 for Pietro's reputation as an astrologer and magician in the Renaissance. According to Chandelier 2013, Pietro's immediate reputation among other physicians in the fourteenth century was not great, perhaps because his thought was so unusual.

16 Pietro is prominent in many books about learned medicine of the late Middle Ages, including Siraisi 1981 and 1990, Ottosson 1984, and Agrimi and Crisciani 1988.

17 Morpurgo 2008, Touwaide 2008, and Ventura 2016.

18 His use of astrology went beyond that of most of his contemporaries, but due to the influence of Pietro among others, astrology became increasingly important in medicine throughout the fourteenth and fifteenth centuries. See Vescovini, 1985, 1987, and 1997; Siraisi 1990: 67–68, Jacquart 1993.

19 See Chapter 5.

20 On Pietro's translating activity, see Thorndike 1942 and 1944, d'Alverny 1985, De Leemans 2008, and Fortuna 2008.

21 *Expositio*, Prologue (ed. De Leemans 2016: 45).

22 On aspects of Pietro's ideas that have caught the interests of modern philosophers, see Nardi 1912/1958 and 1920–1921/1958, Alessio 1976, Hasse 2001, and Robert 2017. Beyond these few articles, Pietro almost never appears in general histories of medieval philosophy, except in passing reference. He is mentioned exactly once in each of *The Cambridge History of Renaissance Philosophy* (ed. Schmitt and Skinner 1988), *The Cambridge History of Medieval Philosophy* (ed. Pasnau 2010), and *The Oxford Handbook of Medieval Philosophy* (ed. Marenbon 2012). He is not mentioned at all in *The Cambridge History of Later Medieval Philosophy* (ed. Kretzmann, Kenny, and Pinborg 1982).

23 On the idea of "Latin Averroism", see Marenbon 2007. The idea of Pietro as the founder of "Paduan Averroism" originated in the nineteenth century with Ernest Renan (see Renan 1852: 258–259).

24 See Randall 1940/1961, Wallace, 1995, and Ottaviani 2004. For a summary of earlier views, see Ottoson 1984: 98–126.

25 On the lack of modern respect for eclecticism, see Kelly 2001.

26 See d'Alverny 1966.

prologue to his commentary on the *Problemata* where he claims that, "this book cannot be fully understood except by someone who has studied every part of philosophy"; this, he adds, is why no one has commented on the text before him.[27]

Taking Pietro's thought seriously today helps us recover the authentic character of philosophical and naturalistic ambitions of the thirteenth and fourteenth centuries.[28] This is particularly visible in how we understand "Aristotelian" thought of this period. As for most other medieval scholars, Aristotle was Pietro's pre-eminent philosophical authority, however open he may have been to a variety of intellectual influences. When he prefers a medical or astrological doctrine that appears to contradict Aristotle, Pietro had a variety of tools to explain this, but he rarely indicates that his ideas might be opposed to the Philosopher in any fundamental way.[29] For this reason, some of Abano's more materialistic emphases can sound strange to a reader familiar with modern interpretations of Aristotle's texts or with the descriptions of medieval Aristotelianism usually found in modern accounts.[30] In his commentary on the Pseudo-Aristotelian *Problemata*, especially on the problems concerning moral virtues, his reading can seem quite far from Aristotle.[31] Yet he seems to have considered his medical anthropology to be perfectly consistent with Aristotle.

We should recognize that there is no uniform set of "Aristotelian" doctrines in the Middle Ages. Within the diversity of the Aristotelian thought, it is useful to envision two broad versions of "Aristotelianism" in this time: one that was more metaphysical, and more familiar to modern students of philosophy, and another with a more materialistic and naturalistic bent.[32] This is hardly surprising given the complexity and diversity of the Aristotelian corpus.[33] Within his psychological and biological works in particular, there is significant room for interpretation about the relative priority of

27 *Expositio*, Prologue (ed. De Leemans 2016: 49): "Ex quo quidem deducitur quod liber hic non potest plene intelligi nisi ab illo qui philosophiam secundum omnem partem eius inspexerit; propter quod fortasse in eius expositione pigritarunt glosatores."

28 As Vasoli 1974: 413–417 suggests.

29 See Chapter 5 on incantations for one of these rare examples.

30 Which is not to say that Pietro was a true materialist, but just that he pays great attention to corporeal causes. For an examination of his materialism, see Robert 2017.

31 On this commentary and Pietro's defense of the authentic Aristotelian authorship, see Chapter 6. For a general description of the *Problemata* and the *Expositio*, see below: 11–13.

32 On the notion of "Aristotelianisms", see Schmitt 1983, Grant 1987, and Thijssen 1991.

33 Different Aristotelian works emphasize material or physiological causes to greater or lesser degrees. The texts that were central to the education of theologians and metaphysicians, included the *Metaphysica*, the logical works, the *Ethica*, and *De anima*. More materialistic texts include *De animalibus, Meteorologica, Parva naturalia*, along with Pseudo-Aristotelian works such as *Problemata, Secretum secretorum*, and *Physiognomia*.

the material or formal descriptions of the soul and its functions.[34] The interests of the theologians at Paris and Oxford, where Aristotle was taught as a preliminary to the study of theology, led to an emphasis on the texts and ideas that tended to accentuate immutable truths, perfection, and transcendence of our material existence. Even more physiological texts tended to be "theologized".[35] Meanwhile, the texts central to the medical curriculum placed greater emphasis on physiology; the central Aristotelian texts here suggest gradualism between humans, animals, and the natural world and common physiological explanations. Finally, if we include pseudo-Aristotelian texts, such as the *Secretum secretorum* and the *Problemata* in the authentic Aristotelian corpus, as Pietro and his contemporaries certainly did, Aristotle looks even more physiological. Pietro's own Aristotelianism is more recognizable in these texts of the "physiological" Aristotle. It is further shaped by his willingness to incorporate outside elements from an eclectic variety of allied disciplines. All of this enriches our understanding of the definition of philosophy in this period.

There are a couple of general features of Pietro's thought that deserve special mention. The first of these relates to his conviction in the causal power of astral forces, certainly the most important aspect of his thought not treated in depth in this study, because it is amply treated elsewhere.[36] The importance of astrology in medicine was based on the widely held belief among naturalists and inherited from Arabic sources that the celestial bodies functioned as causes of events on earth, both as remote causes that set more immediate terrestrial causes in motion and, sometimes, as immediate causes in themselves. Thus, for the physicians who approved of astrology, this discipline was regarded as obligatory for understanding the causes of disease. Not all physicians considered astrology to be important. For this period, the seriousness with which astrology was regarded can be taken as an indicator of the philosophical ambition of individual physicians. For Pietro, astrology became a way to refute the classification of medicine as an art by tying it to eternal movements. To critics who maintained that medicine dealt only with contingent events which could not be reduced to science, Pietro and others countered that the eternal and regular motion of the spheres also assured that there was an inherent rationality in the world below and that this rationality could be discovered, even if it was not immediately apparent.

For Pietro, knowledge of the celestial bodies represented the highest knowledge possible and provided medicine with a scientific foundation.[37] As part of the foundation of Pietro's medical philosophy, the motions of the spheres gain in importance

---

34 On the tension within Aristotle's biological works about the priority of form (soul) and matter, see Lloyd 1996: 38–66. For Pietro's almost physicalist account of the introduction of the soul, one of the main loci for this tension, see Chapter 3.

35 French and Cunningham 1996.

36 Among the many works that consider Pietro's astrology specifically, see Vescovini 1987 and Jacquart 1993.

37 Vescovini 1987.

when one considers his simultaneous rejection of divine things and separated intelligences as objects of scientific knowledge.[38] This rejection excluded from consideration God and the "intelligences" that many other scholastics located in the stars. It is the observable motions of the celestial bodies and the effects they produce on earth, not divine intelligences in the stars and planets that we should seek to understand.

Pietro's commitment to the necessity of astrology in medicine is indisputable. He devoted his *Lucidator* and *De motu octavae spherae* exclusively to astrology and he also translated some astrological treatises by the Jewish philosopher Abraham ibn Ezra (1092/1093–1167) from French into Latin.[39] More important for the claim for the use of astrology in medicine, he included a question about astrology among the ten introductory questions in the *Conciliator*. Asking, "Whether any existing *medicus* is able to heal the sick through the science of the stars?" Pietro expresses confidence that this will one day be true.[40] There is no doubt that Pietro saw astrology as a vital part of the natural concerns of physicians and an important element in his inclusive understanding of medical expertise. The defence of astrology by Pietro and by Arabic physicians before him came in spite of the fact that neither Aristotle nor Galen (129–216/217 CE) approved of the subject. However great Galen's authority might have been, medieval medical philosophy was essentially eclectic. Most relevant to this current study, the causal connection Pietro established between the apparent contingency of terrestrial events and the necessary motions of the heavens enabled some of his bold pronouncements about the ability of medicine to speak authoritatively about human nature and other topics.[41]

The power of the celestial bodies to affect everything in the terrestrial realm, including human souls and the course of history, were probably some of his doctrines that concerned the Dominicans in Paris. Additionally, he shares with Roger Bacon the conviction that forces or "intentional species" emanate from a whole multiplicity of natural bodies; those emitted by the stars are simply the most powerful.[42] Such forces are quasi-material, possessing varying levels of materiality or spirituality depending on their source. Whether and how they affect other natural objects in their surroundings depends on their own qualitative make-up and on that of their recipient. Light is a visible example of these forces and provides us with a model for how they radiate,

**38** Because our knowledge of God and the intelligences has no basis in sensitive cognition, "they transcend the power of our intellect, since it cannot understand without phantasms", *Lucidator* (ed. Vescovini 1988: 135). For a similar sentiment in the *Conciliator*, see diff. 3 (f. 7r, col. 1, D).

**39** On the *Lucidator*, Vescovini 1988. On *De motu octavae spherae*, see Vescovini 1987: 32–34. On this translation, see Thorndike 1944b and Sela 2019.

**40** *Conciliator*, diff. 10 (f. 15v–18r).

**41** In this, I agree with the argument of Vescovini 1997.

**42** On Bacon's explanatory system, see Lindberg 1983: lii–lxxi, and Tachau 1988: 3–26.

but most are "occult", or hidden, meaning they affect our interior senses directly, and from there further affect the body.[43]

Another constant feature of Pietro's work is his self-conscious and sometimes strident rhetoric about the appropriateness of his naturalistic approach.[44] He writes that when inquiring into natural matters, we must leave out the divine and miraculous, and seek the truth in the words of the astrologers, philosophers, and physicians.[45] Physicians above all are *diligentes ministri naturae* "diligent ministers of nature": they must understand the way nature works so that everything does not appear to happen by fortune or divine impulse, and must harness this knowledge to maintain health and cure illness.[46] Pietro is a consistent critic of those who would prioritize supernatural explanations to things; in this, he is not referring to unlearned people, but to his colleagues in theology, disdainfully referring to those "*theologizantes*", "*Cristocoli*", or "*divini hypocriti*", whose arrogance makes them believe they can know things about God and the intelligences that are beyond the ability of the human intellect.[47] His personality comes through quite forcefully in such polemic. It is easy to imagine him, as a physician first trained in the more secular climate of the Italian schools, rolling his eyes at the vigorous debates about divine illumination and transcendent human perfection that transpired during his time at Paris.[48]

Neither his derisive remarks about colleagues, nor the fact that he confined his research to the realm of nature, much limited his assessment of his own expertise. As has been observed, he was willing to offer natural explanations for topics usually left to theologians. Nor did he back away from even the most controversial phenomena that he felt could be explained by natural causes,[49] including all aspects of sexuality,[50]

---

**43** For further discussion see Chapters 3 and 5.

**44** For the emphasis on Pietro's "naturalism", see d'Alverny 1966, Alessio 1976, and Vescovini 1997.

**45** *Conciliator*, diff. 9 (f. 14v, col. 2, G–H): "Vero sciendum, quod cum de naturalibus differatur secundum Albertum in primo *De Generatione*, nihil ad nos de miraculis divinis, nec etiam legum persuasionibus magis obsequentes, quod multoties apparet, quam quae causarum causa voluntate quadam produxit antecedente. Nil ad praesens de miraculis ipsius et voluntate scrutandum consequenti, cum leges vim solam obtineant persuadendi (*Poli* 2). Inquiratur igitur veritas quaesiti ex dictis astrologorum, philosophorum, et medicorum deinceps."

**46** *Conciliator*, diff. 7 (f. 10v, col. 2, F).

**47** See, for example, *Conciliator*, diff. 135 (ed. Boughan 2006: 299) and *Lucidator*, diff. 1 (ed. Vescovini 1988: 125).

**48** See the end of Chapter 1: 41–44.

**49** Jacquart 2001: 40–41, observes that Pietro did not seem to let the condemnations of 1277 affect his positions.

**50** See Cadden 2013 and 2016, and Coucke 2016.

and same-sex relations in particular,[51] miracles,[52] transubstantiation,[53] the nature of Jesus,[54] and, of course, the subject of this book, his medical anthropology.

## Pietro's Works

This book concentrates on Pietro's major texts, the *Conciliator* and the *Expositio Problemata*, with some use of the *Compilatio Physionomie*.[55] Like Pietro's other writings, which include original treatises in medicine and astrology, commentaries and additions, and translations, these texts are somewhat idiosyncratic. Pietro commented on none of the standard works of medicine or philosophy and left no treatises on typical subjects. At the same time, even in the realm of scholastic literature, his writing can be dense and difficult to follow. This situation makes his work hard to compare directly with that of his contemporaries on specific doctrinal points. Because of these difficulties it will be useful to briefly discuss the structure of the relevant works here.

Pietro's *Conciliator* (1310), the work for which he is best known, is an encyclopaedic medical work comprised of 210 questions called *differentiae*. He tells us in the prologue that he worked on this book for ten years (*iam binis lustris*) before he published it, so it probably derived from his teaching at Paris.[56] Its title reflects Pietro's intention to reconcile differing authorities, especially in medicine and philosophy, which also included astrology, but the work is certainly medical in its structure. The organization of the *Conciliator* is derived from the traditional framework for medical knowledge first created at the medico-philosophical school of Alexandria in Late Antiquity and represented in the most popular later medical textbooks available in the thirteenth century, including the *Isagoge* of Hunayn ibn Ishāq (ca. 809–873) or "Johannitius" in Latin; the *Pantegni* by 'Ali ibn al-'Abbās al-Magusi (d. 994), Haly Abbas in Latin; and the *Canon* of ibn Sina, or Avicenna (980–1037).[57] At the same time, Pietro's selection of material was very much guided by his own interests. He begins with ten questions about the nature of medicine as a discipline, including the object of medicine and medical methodology, situating medicine in broad intellectual perspective. This constitutes much more introductory, and theoretical, material than was present in his models; most of these ten questions are closely related instead to Galen's more free-flowing introductory text known in the

51 On Pietro's analysis of same-sex relations, see Cadden, 2013.

52 Thomas of Strasburg (d. 1357) reports that Pietro had derided divine miracles by giving a naturalistic explanation of Lazarus being raised from the dead. Essentially, Lazarus had an illness that made him appear to be dead for three days. See Nardi 1920–1921/1958: 21–23, for Thomas of Strasburg's report.

53 *Conciliator*, diff. 156, ed. Delaurenti 2012: 77. On this see Chapter 5: 102.

54 *Conciliator*, diff. 18 (f. 28v, col. 2, G). See my discussion in Chapter 2:60.

55 A complete list of Pietro's works with lists of manuscripts can be found in Paschetto 1984.

56 *Conciliator*, Prologue (f. 1r, col. 1, D).

57 On these texts and the structure of medical encyclopedias, see Chapter 1: 18–21.

Late Middle Ages through its Latin translation entitled *Tegni*. Others, such as *diff.* 9: "Whether human nature has deteriorated from what it was in antiquity?" and *diff.* 10: "Whether any existing *medicus* is able to confer health to the sick through the science of the stars?" reflect Pietro's particular interests.

The next 200 questions are divided equally between the theory and practice of medicine. The 100 questions about theory mostly progress in a familiar order: first are the seven "naturals" (*diff.* 11–65), i.e. elements, complexions, humors, parts of the body, virtues, spirits, and operations. These are followed by the "non-naturals" (*diff.* 66–71) and the "*praeter* naturals" (*diff.* 72–77). He finishes the theoretical section with questions about signs (*diff.* 73–86), fevers and a few other illnesses (*diff.* 87–102), and then crises and critical days (*diff.* 103–110).

The final 100 questions about practice are divided into the conservation of health (*diff.* 111–131) and the cure of illness (*diff.* 132–210). The cure of illness has the following order: four general questions about regimens of curing (*diff.* 133–136), several about *materia medica* (natural products for medicines) (*diff.* 136–152), curing in general (*diff.* 153–157), digestion (*diff.* 158–160), evacuation (*diff.* 161–168), cure of fevers (*diff.* 169–173), cure of other illnesses (*diff.* 174–180), questions about illnesses afflicting specific parts of the body, organized from head to toe (*diff.* 181–206), closing with four practical questions not fitting into other categories (*diff.* 207–210).

Each *differentia* begins with a question, e.g. *diff.* 17: "Whether complexion is a substance or not?" After posing the question, Pietro lists citations from authorities against the opinion he will ultimately defend; this is followed by a heading called *Oppositum*, where he briefly gives evidence from authorities in opposition to the first set of opinions. The next four parts of the *differentia* are labelled *Propter* (or "regarding") 1–4. In *Propter* 1, Pietro defines the terms in the question. If these have been previously defined, he gives a reference. *Propter* 2 is a review of authorities' opinions on the question. *Propter* 3 is Pietro's own opinion and is usually the longest section. In *Propter* 4, Pietro responds to the original evidence against his conclusion. Occasionally, issues will come up in the middle of a *differentia* that cause Pietro to append one or more unnumbered questions.[58] He approaches each of these sub-questions with the same thoroughness as the original *differentia*.

The *Conciliator* survives in numerous manuscripts, and it was printed at least 18 times into the seventeenth century. The text of the *Conciliator* quoted in this book is from the reprinted Venice edition of 1565, except for the text of *differentiae* 135 and 156, which comes from the editions of Boughan 2006 and Delaurenti 2012, respectively.

58 E.g. *differentia* 29 has five other questions appended to it, the most for any single *differentia*. It seems that most of these questions were appended, and not numbered, in order to maintain the perfect balance between the theoretical and practical parts of the *Conciliator*.

The *Expositio Problematum Aristotelis* is a commentary on the pseudo-Aristotelian *Problemata.*[59] It is equal in length to the *Conciliator* and written over the same period: Pietro began it at Paris, probably in the 1290s, and completed the final edition in Padua by 1310. The fact that it was the first Latin commentary on this text made it highly influential for the reception and interpretation of the *Problemata.*[60]

The *Problemata* itself is a collection of questions divided among 38 sections (called *particulae* in the Latin translation) on a wide variety of subjects in medicine and natural philosophy.[61] Topics range from hot water, fruits, and barley to the parts of the body, to activities such as sweating, sexual intercourse, drunkenness and moral virtue. The only apparent organization to the collection is that the questions within each *particula* are more or less related to each other; otherwise, they seem quite random and repetitive.[62] The variety of answers to each question reflects the origin of the text as derived from notebooks recording the diversity of solutions proposed within the Peripatetic school. Most scholars today believe that the questions probably developed from an authentic Aristotelian core and were rearranged and augmented over the course of several centuries, probably reaching their final state of revision in the fifth or sixth century.[63] There were apparently already doubts about the Aristotelian authorship of these questions in the thirteenth century, as Pietro defends their authenticity in his prologue.[64]

Problems are a form of question literature that differs from other forms through their emphasis on the paradoxical nature of the question. The starting point for each question, as Pietro explains it, is "what is almost commonly knowable, cast into

---

**59** Pietro's *Expositio* survives in at least 20 manuscripts (including some with Jean de Jandun's minor revisions) and is printed in several early editions, which are listed in Blair 2000. On Pietro's commentary, see: Siraisi 1970, Burnett 1997, Vescovini 2006, Cadden 2013, and De Leemans and Hoenen, 2016. Editions have been made of the Preface (De Leemans 2016), Book 4 (Coucke 2008), Book 7 (Boughan 2006), Book 19 (Meyer 2022), and Book 32 (Devriese 2013).

**60** Although the *Problemata* circulated in Paris soon after their translation into Latin c. 1260, their text remained without a commentary until Pietro's *Expositio.* Pietro's commentary was minimally altered in a version created by the controversial Aristotelian philosopher, Jean of Jandun (1280s–1328) after 1315. On the reception of the *Problemata* and the influence of Pietro's *Expositio* on later commentaries, see the essays in De Leemans and Goyens 2006, and De Leemans and Hoenen 2016.

**61** Aristotle, *Problemata* (ed. and French transl. Louis 1991; English transl. Hett. 1926). On the distinction between this *Problemata* collection and others which also circulated under the name of Aristotle, see Blair 2000.

**62** Each *particula* has a brief heading announcing its contents. These headings circulated with the Greek text and were included in the Latin translation.

**63** Blair 2000 includes the relevant references concerning the original composition of the text. Confusing the matter is that Aristotle, in his authentic works, refers several times to a collection of problems. These references do not correspond to the *Problems* as they have been handed down to us.

**64** See Chapter 6: 111–113 on Pietro's defence of the Aristotelian authorship.

doubt".[65] It "is a question containing some difficulty which must be solved through disputation – something which seems to present an abyss – in that it devours and consumes whatever is proposed."[66] Each problem typically begins with an observation of a phenomenon, which is never itself called into doubt, and asks why it happens. The text then usually suggests possible causes for the phenomenon, presenting these tentatively in the form of questions. Sometimes the suggestions lead to other questions or digressions.

For his commentary, Pietro relied on the Latin translation of the Greek text made by Bartholomew of Messina ca. 1260.[67] This translation proved difficult to decipher for a number of reasons, not least because of Bartholomew's word for word translation style of a text that was already quite difficult and abbreviated in the Greek.[68] Because the subject matter is constantly changing, it is usually impossible to rely on context to sort through difficult passages. To make matters worse, by the time Pietro used Bartholomew's translation, errors within the manuscript tradition had rendered some questions almost nonsensical.

Pietro supplies a regular structure to his exposition of each problem, identifying and commenting on each element of the text. He always begins by rephrasing the ques-

**65** *Expositio*, Prologue (ed. De Leemans 2016: 47): "Materia sive subiectum huius libri vel scientie est scibile pene commune sub dubitatione prolatum."

**66** *Expositio*, Prologue (ed. De Leemans 2016: 49): "Problema quidem est grecum latine prolatione importans. Est enim questio difficilis aliquid continens quod disputatione solvendum est; quod et voraginem videtur denotare – ut enim vorat et accipiat propositum – est problema."

**67** The text translated by Bartholomew actually ended in the middle of book 37 of 38. Pietro did not realize that he did not possess the entire text; however, he was aware that there were supposed to be 38 books, so he divided book 37 into two short parts.

**68** Until recently, it has been assumed that Pietro translated the *Problemata* himself, perhaps being unaware of, or dissatisfied with, the earlier translation of Bartholomew. See De Leemans 2007 on this. This assessment was based on several references Pietro makes to his own translation of a book of problems "attributed" to Aristotle that he says he found in Constantinople (e.g. *Conciliator*, diff. 1 [f. 3r, col. 1, A]: "*Problematibus* Aristotelis attributis per me translatis"). However, Pietro states in the prologue that his *Expositio* is of the authentic work of Aristotle, so it seems unlikely that he would refer to the same collection as merely "attributed to Aristotle", especially when he was aware of the existence of other collections of problems, including the *Problems* of Alexander of Aphrodisias, which he says he has translated. More significantly, Pietro shows no signs of having seen the Greek text. If he *had* translated the Aristotelian *Problemata* himself, and nevertheless decided to base his commentary on Bartholomew (e.g. because it was better known), this would have placed him in a unique position to critique Bartholomew's translation. To the contrary, his commentary adheres to the translation of Bartholomew as he found it in the manuscripts, mistakes and all. Thus, it can be said with some certainty that he not only did not translate the *Problemata*, but that he never saw their Greek text. For a comparison of Pietro's paraphrases of the text to variant readings which occur in the transmission of Bartholomew's translation, see Coucke and Swaenepoel 2009.

tion, sometimes expanding or clarifying it.[69] If possible, he identifies the *motivum* (motive) for the question – i.e. the paradoxical aspect of the question which makes it a true problem.[70] For the remainder of the text he first refers to the lemma from Bartholomew's translation, again paraphrasing it, and classifies each passage as a *signum* (sign) or *causa* (cause), numbering the causes as he progresses through the text. He carefully distinguishes the different types of evidence presented, distinguishing merely probable causes from the more important true natural causes. Throughout, he defines difficult terms used in the question, qualifies the information offered in the base-text, anticipates and resolves doubts about the text (usually introduced by *Dubitur*), supplies additional information, and points out disagreements based on other authorities or his own experience (introduced with *Nosce* or *Dicendum*).

The frequent doubts, digressions, and clarifications are particularly useful for identifying Pietro's own point of view in the commentary. Through these and more subtle means, he steers the interpretation in a specific direction. The numerous instances where Pietro misinterprets the text are instructive as well, because they sometimes reveal his biases and expectations to a greater extent than when he simply repeats the suggestions made in the text.[71]

Pietro's commentary on the *Problemata* did not initiate a new trend. There are very few commentaries on this text written after him and most of these are heavily dependent on his work. Even after the completion of the *Expositio* in 1310, it remained outside of the regular scholastic curriculum. Probably its miscellaneous composition did not lend itself to classroom teaching. As Pietro describes it, unlike the rest of the Aristotelian corpus, the *Problemata* do not belong to any single branch of philosophy, but rather contain discourses on every field, with the exception of grammar and metaphysics.

The text of *particulae* 27–29 used in this book is the provisional text for an edition of those sections in preparation by myself and the late Pieter De Leemans. Text cited from other sections is taken from the first edition (Mantua, 1475), with corrections based on the consultation of MSS Cesena, Biglioteca Malatesiana, XXIV.Dext.2; Nürnberg, Staats Bibliothek, Cent.III.38; and Paris Bibliotheque nationale de France, latin 6541. Because the 1475 edition has no folio numbers, references will cite the *particula* and problem number (e.g. 28.7).

---

**69** E.g., *Problemata* 27.9: "Propter quid timore tristicia quadam existente et dolore? Dolentes quidem clamant; timentes vero tacent." This becomes in Pietro's *Expositio*: "Quare cum timor sit ut quedam tristicia et dolor similiter – et id est motivum ipsius – qui dolent clamant et vociferant sed qui timent silent." (Why, since fear is a type of sadness and pain is also – and this is the motive of the problem – do those in pain cry out, but those who are afraid are silent).

**70** In the case of 27.9 above, the "motive" is that fear and sadness are both species of pain, but they have opposite effects, namely people cry out in pain, but not in fear.

**71** E.g., *Problemata* 29.10.

Pietro's *Compilatio Physionomiae* or *Compilation of Physiognomy* (1295) is a summary of tendencies associated with various natural dispositions of the body and with celestial forces. He called this treatise a "compilation" because it collects the physionomic knowledge of the Greek and Arabic authors, naming [Ps.]-Aristotle (ca. 300 BCE),[72] Polemon (ca. 88–144),[73] Loxus (Hellenistic period),[74] al-Rāzī (or Rhazes, ca. 864–925 or 935),[75] and Avicenna,[76] and others into one single handbook.[77]

Physiognomy as a science originated in the Aristotelian school and the pseudo-Aristotelian text produced around 300 BCE was probably the earliest physiognomic text and certainly the most influential on the later texts.[78] The goal of physiognomy was to identify natural inclinations by examining the outward appearance of the body. This includes innate signs and habits of the soul that have developed later but have altered the body. Pietro explains in the prologue that, according to Aristotle, "it is possible to know the natures of animals if one grants that the body and soul are transformed together, according to natural passions."[79] He is clear that these natural signs do not necessitate behavior, saying "this philosophy does not assign any necessity at all to the acts and habits of men, but shows certain inclinations and impulses in the blood and spirits."[80] In his treatise, he arranged this knowledge as a long list of physical characteristics together with the virtues of the body and soul that they signified.

After an introduction where he explains the principles of the science and characteristic differences between males and females, he divided the rest of the treatise into two longer books, one on the body and one on the effects of astrology. Most of the information is very brief. We learn that people who have a "very light black" (*niger valde levis*) skin color tend to be timid and stupid,[81] while people whose skin is "red splotched with white" (*rubeus albo conspersus*) are brave, as are people with hairy backs.[82] A high hair line corresponds with a small intellect.[83] Ears that are narrow

72 For a Greek edition and English translation of the Aristotelian *Physiognomy*, see Swain 2007: 637–661. For the Latin text translated by Bartholomew of Messina, see Foerster 1893: 1.2–91.

73 For an Arabic and Latin edition of Polemon, see Hoffman ed. in Foerster 1893.

74 For the identity of the mysterious Loxos or Loxius, see Boys-Stones 2007: 59–64.

75 For an edition of the Latin translation by Gerard of Cremona in 1175, see Foerster 1893: 2.161–179.

76 No physiognomy by Avicenna has been identified. See Ghersetti 2007: 293.

77 I use the edition of Padua, 1474, with reference to MS Vatican, *Barberinianus latinus* 341. This edition is not paginated, so numeration is based on my own count.

78 On the origins of physiognomy, see Touwaide 2024, which includes up to date literature. On the pseudo-Aristotelian text, see Boys-Stones 2007: 19–59.

79 *Compilatio*: 3: "[E]docuit Aristoteles naturas animalium cognoscere est possibile si quis concedit corpus simulque animam transmutari secundum ea quecunque naturales passiones."

80 *Compilatio*: 4: "[H]ec philosophia actibus et moribus hominum necessitatem minime impinget, sed inclinations ex sanguine spiritibusque quasdam impulsivas ostendit."

81 *Compliatio*: 9.

82 *Compilatio*: 15.

83 *Compilatio*: 14.

and oblong reveal an envious character.[84] The bulk of the attention is given to the parts of the face and especially the eyes, which are the "doors to the soul" (*oculi velut fores anime*), according to Polemon.[85] The color, shape, size, quality, position, and motion of the eyes are all treated extensively as signs of the soul.

Pietro reminds his readers that physiognomy is far from a demonstrative science because the information learned from physiognomy is ambiguous, in that it doesn't have the status of being true by necessity. To the modern reader, this collection of information looks eclectic, aphoristic, and obviously a pseudo-science, even in comparison to other medieval sciences. To Pietro, it constituted one part of a larger system of connections between the body and soul.

**84** Notice the exaggerated oblong ears on the painting of the personification of "*invidia*" by Giotto on the Scrovegni Chapel in Padua. Giotto painted this while Pietro was in Padua and the first reference to Giotto occurs in Pietro's *Expositio* 34.1. See Thomann 1991.
**85** *Compilatio*: 32.

# Chapter 1
# Body-Soul Relations in Medical Philosophy from Salerno to Pietro d'Abano

For most scholars in the Middle Ages, medicine was a mechanical art, concerned with the practical business of restoring and maintaining physical health rather than a science producing theoretical knowledge about the natural world or the human body.[1] This practical or therapeutic side of medicine certainly describes European medicine from about 600–1050, when the surviving literature seems to reveal little interest in theorizing about medical advice.[2] However, an alternative vision of medicine as a "second philosophy" also survived in this period, most noticeably in the popular *Etymologies* of Isidore of Seville (before 570–636).[3] This philosophical side of medicine became more prominent beginning in the late eleventh century, when a wave of translations of Greek and Arabic medical texts marked a theoretical turn for scholarly medicine, communicating the idea of medicine as both theory and practice.[4] The three parts of theory – physiology, pathology, and semiology – offered a systematic framework for understanding the human body and its interactions with the natural world. It also marked the beginning of medieval discussion about the ability of medicine to contribute to a host of questions about human nature and morality. The role of medicine in philosophical discussion diminished in the early thirteenth century, as medical teaching was established in the new universities and Aristotelian natural philosophy became dominant. By the end of the thirteenth century, after much of Galenic medicine had been translated and assimilated, Pietro d'Abano and several other physicians began to take seriously Galen's claim that not only was theory important for medicine, but the best physician is also a philosopher.[5] Study of the full range of Galen and Arabic Galenism led to robust discussion of the authority of medicine to

---

1 On the classification of medicine within the arts and sciences, see Weisheipl 1978: 472–473.

2 For medical texts available in Europe in the early Middle Ages ($9^{th}$–$11^{th}$ centuries), see Beccaria 1956. For a summary, see Sigerist 1958. On interpreting some early medieval texts, see Wallis 1995 and 2000. On the interpretation of an early medieval teaching text in Salerno, see Glaze 2011. For a comparatively positive assessment of this period, see Horden 2009.

3 *Etymologies* IV.13, ed. Lindsey 1911 (Engl. transl. Barney et al. 2006: 115). On Isidore and his influence, see Fear and Wood 2019.

4 For the division of medicine into theory and practice, see Demaitre 1975: 103–123, Ottosson 1984: 65–97, and Agrimi and Crisciani 1988: 21–47. This division is a late Alexandrian development and remained a central feature of Galenic medicine thereafter.

5 For Galen's short treatise of this name, see Galen, *Selected Works*, transl. Singer 1997: 30–34. For a discussion of how the relationship between medicine and philosophy shaped the interpretation of Galenic thought in Alexandria and the Arabic world, see Temkin 1973: 61–80. For recent work on Galen and his reception, see Bouras-Vallianatos and Zipser 2019, and Nutton 2020.

https://doi.org/10.1515/9783110612639-002

speak to current philosophical concerns and laid the groundwork for Pietro's novel medical philosophy.

## The Long Twelfth Century

### New Translations and Commentaries

As part of the larger cultural movement known as the "Twelfth-Century Renaissance", a number of translations of Arabic and Greek medical texts changed the character of medical knowledge in Western Europe.[6] The earliest and most important sites for the production of the new medical texts were the monastery of Monte Cassino and the nearby "school" in Salerno where archbishop Alfanus of Salerno (d. ca. 1085) and Constantine the African (d. ca. 1087) were the central figures.[7] Sponsored by Alfanus, Constantine translated and adapted some of the Hippocratic and Galenic literature produced in Alexandria and Baghdad from the eighth to tenth centuries.[8] Constantine did not produce literal translations and, although he knew little or no Greek, he gave his texts a Greek guise, sometimes supplying Hellenised names, and removing all references to Arabic works and authors.[9] If he intended to expedite the reception of these texts, he was extremely successful. Among his many translations were the texts that would form the core of the medical curriculum into the sixteenth century. These included the original treatises of the collection that would come to be known as the "*ars medicine*" or "*articella*", which served as a kind of general introduction to medicine.[10] Most of these and other translations produced by Constantine's circle were practical in orientation, but two texts in particular provided the theoretical framework for medicine, the *Isagoge*, a brief introduction to Galenic theory, and a longer encyclopaedic work, the *Pantegni*.

---

6 For an overview of twelfth-century translating activity see d'Alverny 1982, and Burnett 2013a. For an account of the "Twelfth-Century Renaissance", which incorporates medical learning, see Burnett 2013b.

7 On the beginnings of theoretical medicine in Salerno, see Kristeller 1945/1956 and 1976, Garcia-Ballester 1995, Jacquart and Paravicini Bagliani 2007, and Glaze 2019. Salerno is now regarded less as a formal institutionalized "school" and more as a school of thought. See Ziegler 2008.

8 For Constantine and his work, see Burnett and Jacquart 1994, and Glaze 2019. For a summary of Constantine's translations, see Jacquart 1993/1998: 202–205.

9 On Constantine's Hellenizing, see Jacquart 1986 and 1994, and Gasper and Wallis 2004: 158–160.

10 The *articella* originally consisted of the Hippocratic *Aphorisms* and *Prognostics*, Theophilus on *Urines*, Philaretus on *Pulses*, and the *Isagoge ad artem Galeni*. Galen's *Tegni* was added by the mid-twelfth century. The origin and development of the *articella* remains in doubt. See Kristeller, 1976: 64–71, and Gasper and Wallis 2004, 149–163. For the later thirteenth-century development of the *articella*, see O'Boyle 1998.

The *Isagoge* is a heavily abridged version of a ninth-century Arabic work by Hunayn ibn Ishāq, a Nestorian Christian working in Baghdad, which Constantine rendered into Latin with the Hellenized name "Johannitius".[11] Hunain's work was itself an adaptation of fifth- and sixth-century Alexandrian introductions to Galen, so the *Isagoge* transmitted Alexandrian Galenism in abbreviated form with some revisions.[12] At first glance it looks rather rudimentary, being a schematic introduction and simplification of the various parts of medicine, with cursory descriptions of each term. It was the first Latin text organized with the fundamental division of medicine into theory and practice, with further subdivisions of each part.

Theory starts with a general description of the human organism, beginning with the seven "naturals", which are the categories of things innate to an individual, their balance and proper operation defining health. According to the *Isagoge* (sec. 2–15) the naturals include the four elements, complexions, humors, parts of the body, faculties or powers (*virtutes*), operations (sometimes also rendered "faculties"), and spirits. This terminology echoes philosophical terminology, manifesting the link between medicine and philosophy. In particular, the system of the elements and complexions – the qualitative mixture of hot, cold, wet, and dry arising directly from the mixture of the elements – is derived from Aristotelian biology.[13] The *Isagoge* also uses the Aristotelian notions of "substantial" and "accidental" to classify signs and symptoms of illness.

Of particular interest is the fact that physiology and etiology were applied to the operations of the soul, offering material explanations for psychological phenomena. Most of the psychological material is contained in the categories of the powers, operations, and spirits. The powers are divided into the three systems of the natural, spiritual, and animal, each seated in appropriate parts of the body and dependent on particular physical dispositions (*Isagoge* sec. 12–15). This Galenic tripartite division had the philosophical advantage of being roughly compatible with Platonic and Aristotelian tripartite theories of the soul. The natural powers include our lower functions such as digestion, nutrition, and growth; the spiritual power dilates and contracts the heart and produces emotions; and the animal powers in the brain produce three distinct acts of ordering, volition, and perception. The first of these acts is divided into imagination in the front of the brain, cogitation in the middle, and memory in the back. The "operations" (sec. 16) classify the actions of these powers, divided into simple and complex. Next, there are three types of "spirits" (sec. 17) subtle material flowing through the veins, arteries, and nerves, divided into natural spirits originating in

---

**11** On Constantine's translation, see Jacquart 1986, and Newton 1994. On Hunayn and his translations, see Lamoreaux 2016. On the choices of the names "Johannitius" and "Isagoge", see Jacquart 1986: 231. For the Latin *Isagoge*, see Johannitius, ed. von Maurach 1978. For an English translation, see Wallis 2010: 139–156.

**12** On Hunayn, see Cooper 2011: 13–23.

**13** On complexion, see Chapter 2.

the liver, vital spirits from the heart, and animal spirits from the brain.[14] A whole range of behaviors, from excellent to deviant, could be materially explained by this physiology.

Further discussion about the soul is found among the six "non-naturals" (sec. 27–41), the categories of things that modify the status of the naturals.[15] The non-naturals include airs, food and drink, motion and rest, sleep and waking, evacuation and retention, and the passions (or emotions) of the soul. Sections on the passions are of the most interest for the soul. For the purposes of practical medicine, the passions could be defined in terms of their effect on the normal state of the body and thus their potential to affect health or illness, or the regular or irregular operations of the soul.

The *Pantegni* develops these concepts more thoroughly. It is a loose translation by Constantine of an encyclopaedia composed in Arabic in Baghdad by the tenth-century Persian physician 'Ali ibn al-'Abbās al-Magusi (d. 994), known in Latin as Haly Abbas. It was intended as a comprehensive guide to all medical knowledge, an ambition reflected in the Hellenized name, meaning "the whole art".[16] As in the *Isagoge*, the *Pantegni* divides medicine into theory and practice, progressing in the same manner by first considering theory then practice; however, it has much more complete descriptions of each subject than the *Isagoge*. Within the "naturals" (Books 1–4), the "complexions" (Book 1, Chap. 5–22) and the powers (*virtutes*) (Book 4, Chap 1–17) are far more developed in the *Pantegni*, while the elements (Book 1, Chap. 4), which are generally subsumed within the complexions, and humors, not as important to philosophical medicine, receive little attention (Book 1, Chap. 25).[17] Constantine's text also offered a holistic vision of medicine, not just in the treatment of the broad range of factors affecting health, but by encouraging the study of subjects beyond those of immediate relevance to medical practice. Thus in the theoretical section, the *Pantegni* discusses the variety of skills necessary to be a good physician; these include not only practical skills like dexterity, memory, and sense perception, but knowledge of philosophy and the liberal arts.[18] In an addition of his own to the text, Constantine's preface

14 For an overview of the meaning of "spirit" or *pneuma*, see Llyod 2007, and Chapter 3.

15 On the non–naturals, see Bylebyl 1971, and García-Ballester 1993.

16 On the *Pantegni*, see Burnett and Jacquart 1994, Glaze 2019, and Kwakkel and Newton 2019. Editions of *Pantegni* can be found in *Opera omnia Isaac* 1515: 2.1r–144r. *Constantini Africani . . . opera* 1536/39: 1.324–387, contains books II and IX of the *Practica*, the *De chirurgia* and *De gradibus*; and 2. I–346 contains the entire *Theorica*. On the translation and transmission of the text, see Jordan 1994. For a list of editions and manuscripts, see Burnett and Jacquart 1994: 316–351. Green 2008: 18n7 adds to the list.

17 For an edition of the chapter on the elements from the *Pantegni*, see Burnett 2007. On the spirits in the *Pantegni*, see Burnett 1994, which includes transcriptions of the text of the chapter on the spirits from the two major manuscript families (113–120).

18 *Pantegni* 1515: Chap. 1 f. 1r, col. 2.

included a list of the sixteen books of the Alexandrian Galenic canon, suggesting additional texts for the further development of theoretical medicine.[19]

These two works became the principal sources not only for the development of medical concepts through the twelfth century, but also for speculation about human nature in and outside of medical circles. The doctrines in the theoretical sections present an image of health, illness, and human flourishing governed by the same principles of natural causation that ordered the rest of the natural world.[20]

The other important figure in the early translation of medical theory was Alfanus, who, besides supporting Constantine's Arabic translations, translated several texts from the Greek.[21] His most important translation is the *De natura hominis* or *Premnon physicon*, a Greek work by Nemesius (fl. ca. 400), bishop of Emesa, which aims to expound Christian anthropology with the help of medical theory.[22] Although his view is basically Platonic in that he clearly distinguishes the soul from the body, describing the soul as using the body as a tool, and insisting on its immortality, Nemesius goes on to focus on the body's construction (Sec. 1).[23] Our bodies contain all levels of creation. We are "associated with inanimate things in virtue of the body and the mixture of the four elements, with plants both in virtue of these things and in virtue of the power of growth and generation, and with non-rational beings both in virtue of these things and, for good measure, in virtue of movement by impulse, desire, spirit."[24] Demonstrating detailed knowledge of Galenic anatomy and physiology, Nemesius goes through each of these levels of being in the human body, from elements (sec. 5), to the imagination (sec. 6), sensation (sec. 7–11), thought and memory (sec. 12–13), locating these functions of the soul in the parts of the body, with the highest psychic functions located in the brain, as Galen had demonstrated. Throughout, the emphasis is on the excellence of the human body. Not strictly a medical text – the latter chapters are devoted to arguing against fate on behalf of free will – it presented the possibility of far a more "medical" Christian anthropology than any other text of this period.

By the mid-twelfth century (thus some two generations after Constantine's translations), the earliest commentaries on these theoretical texts emerged at Salerno, Chartres, and perhaps in the Anglo-Norman world.[25] Some of the more ambitious commentators amplified the doctrines related to the soul in the *Isagoge*, compiling opinions from other

---

**19** *Pantegni* 1515: Prologue, f. 1r, col. 1. On the creation of this canon, see Temkin 1973: 69.

**20** Garcia-Ballester 1995 emphasizes the revolutionary nature of the acceptance of the "natural" principles governing the human organism.

**21** On Alfanus, see Gaspar and Wallis 2004: 155–157.

**22** For an English translation, see Sharples and van der Eijk 2008. There is another, later, Latin translation of Nemesius by Burgundio of Pisa (d. 1193) in the edition of Verbeke and Moncho 1975. For the Greek text, see Morani 1987. On the fortuna of Nemesius' text, see Wicher 1986.

**23** On Nemesius' view of human nature, see Dusenbury 2021.

**24** *On the Nature of Man*, 1.2 (Engl. transl. Sharples and van der Eijk 2008: 37).

**25** For this delay of approximately 50 years, see Garcia-Ballester 1995. The centrality of Salerno as the locus of the earliest commentaries on the new translations has been revised since the foundational

sources to explain the psychological material.[26] The first known commentary on the *Isagoge* from the early to mid-twelfth century, the *Digby Commentary* (named for the MS in which it is found: Oxford, Bodleian Library, MS. Digby 108), explains the schematic psychology of the source-text with use of the *Pantegni* and the *Premnon physicon*.[27] The author supplements the outline of the *Isagoge* by adding detail from these sources about the physical locations of psychic powers in the brain and connecting these different powers to the quality and quantity of the humors and elements they require to function properly. The explanations of the soul remain on a very mechanical level, developing the terms in the *Isagoge*. Where the *Isagoge* (sec. 15) gives a minimum of supplementary information about the three powers in the brain ("imagination in the front, cogitation or reason in the cerebrum, memory in the back" *fantasia in fronte, cogitatio vel ratio in cerebro, memoria in occipitio*), the *Digby Commentary* mentions how the nerves link these three powers, what processes they use in their operation (e.g., "distilling" or "digesting" *decoquendo*), and what particular qualities they each require (e.g., the imagination should be hotter than the memory and drier than the cogitation).[28] On the other hand, the *Commentary* does not reveal any awareness that a material description of psychic functions in the brain might conflict with belief in the soul's immortality or preemptively address any potential conflicts. In comparison, the principal sources, the *Premnon physicon*, and to a lesser extent the *Pantegni*, are sensitive to these issues.[29] The commentator's assumption simply seems to be that, for the purposes of a medical analysis, the physiology of the soul does not contradict Christian doctrine. Two other commentaries on the *Articella* from around 1150 by Archimatthaeus (fl. 12th cent.) and Maurus of Salerno (ca. 1130–c. 1214) add information not found in *Digby* with the use of additional non-medical sources.[30] In regard to the sections on psychic functions from the *Isagoge*, the two commentaries add material about the specific elemental qualities needed for various activities. Notably, when he describes the animal powers in the brain, Archimatthaeus contributes a citation from Augustine's *De Genesi ad litteram* locating the principle of operation of the animal powers in the brain.[31] As Jordan points out, "this is by no means a 'solution' to the 'problem' of the soul's connection with a physical body", but it

work of Kristeller, with Morpurgo 1984 and 1990, Burnett 1984, and Green 2008, showing the spread of the texts to Northern France and England.

26 For the psychological focus of some early commentaries on the *Isagoge*, see Jordan 1990 and 1987. Jordan 1990: 45–49 argues a Salernitan origin for the first Latin *Isagoge* commentary, although he considers Chartres a possibility.

27 Jordan 1990: 49–51.

28 Jordan 1990: 52.

29 Jordan 1990: 52–53.

30 On Archimattaeus, see Jordan 1987: 130–131. On Maurus, see Saffron 1972. According to Jordan 1990: 54, new sources include Boethius' commentary on the *Topics*, Galen on the *Aphorisms*, Isaac Israeli's *Universal Diets*, Macrobius, and Aristotle's *Physics*.

31 Augustine, *De Genesi ad litteram* 7.18 (ed. Zycha 1894: 215). Cited in Jordan 1990: 54.

does recognize that this physical explanation is also of philosophical and theological interest.[32]

Two of the last and most famous Salernitan physicians, Bartholomew and Urso, made much stronger claims about the physician as philosopher than their predecessors. Bartholomew of Salerno (fl. after 1150) was probably responsible for including Galen's *Tegni* (*Ars Parva* or *Art of Medicine*) in the *Articella*, inserting it behind the *Isagoge*, and thereby adding a significant amount of theoretical material to these introductory texts.[33] The *Tegni* is organized more loosely than the *Isagoge*, but is generally structured around a trifold division of healthy, sick, and neutral states in bodies, and the relative permanence of these states, with which Galen begins in sections 1 and 2. These three states are then applied to the signs by which they can be determined, moving through the various parts generally from more to less complex (which includes much of the content of the "naturals"), the causes of the states (including the "non-naturals"), and various treatments. It identifies the variety of qualitative mixtures in each part, including many that affect the intelligence, memory, perception, and inclinations, along with physical signs for interpreting the various states. Thus in the brain, intelligence is caused by a fine substance, while slowness indicates a thick substance; good memory indicates a stable substance, and learning quickly a substance that takes impressions easily (but does not necessarily retain them); too much heat is indicated by changeability of opinion.[34] In the heart, courage shows a healthy amount of heat, while "bad temper and crazy rashness" indicate too much heat.[35] In his commentary on the *Tegni*, Bartholomew is quite taken with the possibilities of these ideas, even if there were few other sources available to help with the interpretation of this new material.[36] Faith Wallis argues that Bartholomew marks the "moment when western medicine becomes *physica*, the philosophy of the body, and the physician is transformed from the healer of illnesses into the expert advisor on health and its psychic counterpart, ethics."[37] Since nature and soul are both efficient causes of the body's motion, soul is therefore the concern of the physician.[38] Bartholomew argues that because the soul is conjoined to the body, the physician has a moral role, and that not only could a physician help a patient's moral character, but that a good physician must be morally balanced himself.

---

**32** Jordan 1990: 56.

**33** For the inclusion the *Tegni* within the *Articella*, see Kristeller 1976: 72, and Wallis 2007. For an English translation of the *Tegni*, see Singer 1997: 345–396, and Johnston 2016 (Loeb edition with English translation).

**34** Galen, *Art of Medicine* 6 (Engl. transl. Singer 1997: 353).

**35** Galen, *Art of Medicine* 10 (Engl. transl. Singer 1997: 357).

**36** On Bartholomew, see Wallis 2007, 2008, and 2015.

**37** Wallis 2015: 86.

**38** Wallis 2008: 138–139.

Urso (fl. ca. 1160–ca. 1200) self-consciously styled himself as a true philosopher, calling himself "audacious Urso" (*Ursus audax*) and expressing his desire to not simply learn and comment on *physica*, but to creatively build on his source texts.[39] Rather than commenting on the *Articella*, he left an original treatise on mixtures of elements, one of the central topics in the nexus between medicine and natural philosophy, and wrote a collection of his own *Aphorisms* with his own commentary.[40] According to Maaike van der Lugt, one of these aphorisms (39) contained the first explanation of the healing power of incantations in medieval Europe, a topic that Pietro will later take up.[41] Perhaps building on Bartholomew's claims for the moral status of a good physician, Urso explains that a physician must have an appropriate balance of humors and spirits, which will purify the air around him and enable him to successfully perform incantations.[42] This analysis was indeed more "audacious" than anything we will see from a physician until Pietro d'Abano's generation.

Salernitan doctors also played a major role in the introduction of Aristotelian natural philosophy to medieval Europe.[43] Aristotle (384–322 BCE) is frequently mentioned in the new medical texts as a source, so physicians quickly became aware of the existence of his range of work. The first citations of Aristotle are certainly second-hand (from Abu Ma'shar [787–866] and other Arabic astrological sources), but soon show first-hand knowledge.[44] The later commentaries show progressively greater use of Aristotle's physical works citing the *Physica* and *De generatione et corruptione* and the fourth book of the *Meteorologica* as well the *Categoriae*, *Topica* and logical works.[45] In addition to the content supplied by these texts, the pseudo-Aristotelian *Problemata*, or at least the partial version, known as the *translatio vetustissima*, was extremely popular and provided a methodological model for the genre of Salernitan problem literature.[46]

However, while Aristotle may have enhanced medical theory in the short term, the increased study of Aristotle probably undermined the notion of medicine *as* philosophy in the next century. Aristotle's work would provide more a systematic natural philosophy than the medical theory alone and revealed a very different idea of the relationship between medicine and philosophy than Galenic medicine.[47] Aristotle

39 On Urso, see van der Lugt 2013.

40 Urso, *De commixtionibus elementorum* (*On the Mixture of the Elements*) ed. Stürner 1976, and *Aphorisms*, ed. Creutz 1936.

41 See Chapter 5.

42 van der Lugt 2013: 334–346 contains an edition and English translation of this text.

43 For the role of Salerno in the reception of Aristotle, see Jacquart 1988.

44 For the early use of Aristotle in Salerno, see Birkenmajer 1970, and Jacquart 1988. For Abu Ma'-Shar, see Lemay 1962.

45 Jacquart 1988: 416–423. These texts would continue to be part of the medical curriculum through the following centuries. On the medical uses of the lesser known *Meteorologica*, see Martin, 2002.

46 Lawn 1963.

47 Bylebyl 1990.

often contrasts scientific truth with practical knowledge, using medical examples to clarify just this distinction. Meanwhile, Galen's authentic works, which provided the strongest argument for the *medicus* as philosopher, were rarely used until the last decades of the thirteenth century.[48]

## Philosophical Uses of Medicine

However, before the later dominance of Aristotle, medical theory came to be increasingly identified as *physica* (or natural philosophy) by those outside the medical profession and was sought out by a variety of readers interested in the possibilities of a rationalistic understanding of the cosmos.[49] Anselm of Bec (ca. 1033–1109) in Normandy, requested a copy of the Hippocratic *Aphorisms* between 1070–1077, before he became an Abbot and Archbishop, but little is known about his intentions.[50] Adelard of Bath (ca. 1080–c. 1152), who translated several texts from Arabic himself and was perhaps responsible for bringing several medical and scientific texts from Salerno to France, seems eager to call attention to the novelty of his use of medicine, citing Nemesius, the *Isagoge*, and the *Pantegni* in his *Questiones naturales* (Questions on Natural Philosophy) and other Platonic-style dialogues.[51] Adelard used medicine to elaborate how the body drags down the soul; our material existence explains our current corruption. The bodily passions harass the soul and "reduce man to the level of a beast".[52] Later, in the thirteenth century, Adelard's work would be extremely influential in the thought of Robert Grosseteste (ca. 1168–1253) and Roger Bacon (1214 or 1220–1292), so his work had possibly a greater impact on Pietro's style of thinking than any other twelfth-century philosopher.

The new medical theory figured prominently in the Christianized Platonic cosmological thinking sometimes associated with the cathedral school at Chartres.[53] These Chartrian thinkers were characterized by their conviction that empirical knowledge of the natural world could uncover more transcendent or divine truths.[54] The natural

**48** Green 2019.

**49** Bylebyl 1990.

**50** According to Gaspar and Wallis 2004.

**51** On Adelard, and editions with English translations of *De eodem et diverso, Questiones naturales, and De avibus tractatus* see Burnett 1998.

**52** Adelard of Bath, *Conversations with his nephew* (Engl. transl. Burnett 1998: 33).

**53** On the debate about the unity and existence of the school of Chartres, see Southern 1982. For the general consensus today that it is still useful to conceptualize a unified "school of Chartres", see Wetherbee 1988.

**54** Because of the prominent place of medicine in twelfth-century philosophy, this is a rare time in which medicine is included in modern histories of philosophy, such as *A History of Twelfth-Century Western Philosophy*, ed. Dronke 1988. For an overview of the twelfth-century cosmology, see Wetherbee 1988.

and supernatural orders were conceived as parts of a single coherent order. Belief in the idea that the union of body and soul is a microcosm of the entire universe stimulated interest in the recently translated medical texts.[55] The most influential text in this Platonic cosmology was an incomplete Latin translation and commentary of Plato's *Timaeus* produced by Calcidius (fl. 4$^{th}$ cent.) in the fourth or fifth century.[56] Although Calcidius had not translated the most substantial sections about the human body in the *Timaeus*, he nevertheless clearly indicated the importance of this knowledge in his commentary.[57] At least some twelfth-century readers decided that this missing information about the microcosmic body was knowledge worth pursuing, and the *Pantegni* and the early *Articella* were used in France almost immediately after their translation.[58]

William of Conches (ca. 1085–1154) is an early example of this thinking.[59] William, who probably spent his career at Chartres and Paris, used these works by the 1120s when he composed the *De philosophia mundi*, later revised as the *Dragmaticon* in the 1140s.[60] Always eager to show off his new sources, William cites Constantine and Nemesius frequently, and "pours scorn" on those who ignore the works of the medical authors.[61] William mined the medical concept of the elements in his search for understanding the ultimate constituents of matter and also applied the body's physiology to the macrocosm of the universe (e.g. using physiology to explain the weather).[62]

William was joined by his contemporary, Thierry of Chartres (d. ca. 1150), who similarly uses medicine to elaborate Christian anthropology, including the relationship between the body and soul, the natures of Adam and Eve before the fall, the subsequent effects of original sin on our inclinations, and even explain Jesus' human nature from the perspective of natural philosophy (*secundum physicam*).[63] William, however, after being criticized, backs off on some of his more ambitious claims in his later revision in the *Dragmaticon*.[64]

Bernard Silvester (ca. 1100–c. 1160) shows another variety of the cosmological use of medicine. His *Cosmographia* (late 1140s) is an allegory of the microcosm and macrocosm, created by a variety of personified natural forces.[65] In his poetic account of cre-

---

55 Wetherbee 1988: 25.

56 Edition of Calcidius' commentary (*Timaeus a Calcidio translatus commentarioque instructus*) by Waszink 1962.

57 *Timaeus* ed. Waszink 1962: 256.

58 On the early use of the *Articella* in Chartres, see Burnett 1984.

59 On William, see Elford 1988, and Cadden 1995.

60 William of Conches, *De philosophia mundi*, ed. Maurach with Telle 1980, and *Dragmaticon*, ed. Ronca 1997. See Ronca 1994 on William's use of the *Pantegni*.

61 See Elford 1988: 312, and Cadden 1995: 7.

62 Elford 1988: 321.

63 On Thierry, see Dronke 1988. On Adam and Eve, see Elford 1988: 324–326; and Cadden 1995: 8.

64 Cadden 1995: 4.

65 For an edition and English translation of the *Cosmographia*, see Wetherbee 2015.

ation, after the rest of nature had been created, the first man was fashioned as a model of the cosmos.[66]

The medical category of "spirits" (one of the "naturals") attracted a particularly large amount of attention from philosophers.[67] *Spiritus* has a relatively straightforward status within Galenic physiology: it is an extremely rarefied substance – like a vapor – necessary for the body's functions, but it is certainly material. Vital spirits arise in the heart and arteries, natural spirits in the liver and veins, and animal (or psychic) spirits in the brain and nerves. However, when it was transferred to a theological context in which the word "spirit" had a variety of meanings and was often contrasted with body or equated with soul, the difference between these medical spirits and the immaterial soul took on a certain urgency. The *Pantegni*, after introducing the different types of spirit, comments that "Some wise men . . . have claimed that this spirit which is in the brain is the soul, and that the soul is a body."[68] The text goes on to recommend that this question should be left to the philosophers. A popular philosophical text which took up the issue, was *De differentia anime et spiritus* (On the Difference Between Soul and Spirit) attributed to Qustā ibn Lūqā (or Costa ben Luca, ca. 820–ca. 912), which served as a starting point for further discussion.[69] For Platonic cosmologies, *spiritus* was attractive as an intermediary between body and soul, nature and God or world soul. Building on Qustā ibn Lūqā and his Platonic interpreters, Alfred of Sareshel (fl. late 12$^{th}$–early 13$^{th}$) saw *spiritus* as an almost immaterial, celestial, or super-elemental substance, because it was problematic for a material entity to be the efficient cause of life, or intentionality and other mental activity.[70] According to Bono, Albertus Magnus and other thirteenth-century thinkers would oppose such elevated interpretations of *spiritus*, restoring its medical status as mere – but subtle – matter in the next century.[71]

Although in some ways the second half of the twelfth century was a high point for the use of medicine in philosophy, the prestige of medicine as *physica* was principally related to its use in the search for religious truth from knowledge of the natural world, or for understanding divine things *secundum physicam*, rather than new investigation into medical theory.

---

**66** See Stock 1972.

**67** On "spirits", see Chapter 3. On discourse about spirits in the twelfth century, especially the relationship between the theological and medical definitions, see Bono 1984.

**68** Burnett 1994: 105.

**69** Because Qustā ibn Lūqā was later confused with Constantine the African, the *De differentia anime et spiritus* was printed in collections of Constantine's work. It can be found in Constantinius Africanus, *Opera* 1536: 308–317.

**70** Bono 1984: 98–119.

**71** Bono 1984: 119–122.

## Critics

Moralists, especially North of the Alps, had strong criticism for medicine along with the other kinds of profane knowledge contained in the "new science" of the twelfth century.[72] This hostility was especially directed toward the most audacious uses of medicine that seemed to conflict with Christian doctrine. This criticism entailed both moral and metaphysical arguments: medical therapy required too much attention to bodily desires, and medical theory was based on a vision of nature opposed to religious doctrine. Twelfth-century clergy appear to have been far more hostile to medical speculation than earlier patristic sources, most of which did not regard medicine as inherently dangerous, but only potentially so.[73] Summing up the sentiment of many contemporary moralists, the Parisian theologian Peter Comestor (d. ca. 1178) asserted that the "school of Hippocrates" (*schola Hippocratis*) was irreconcilable with the "school of the Savior" (*schola Salvatoris*).[74] Comestor saw physicians as the holders of knowledge that was not only useless and uncertain in comparison to miracle, but even immoral and dangerous for the health of the soul.

Although the hostility peaked at this time, there had long been a certain ambivalence toward medicine imbedded in Christian rhetoric.[75] Indeed, the lack of theory in early medieval medicine was probably not simply a product of the accidental survival of certain texts. Rather, the theoretical content of ancient medical texts likely had been deliberately excised for some of the same reasons the theory provoked censure in the twelfth century: it was deemed to be at odds with the Christian understanding of the natural world.[76] Copyists divorced particular therapies from their original theoretical context, and connected them to a new foundation, in which causality was ultimately a matter of God's will, while nature acted as just one of several possible intermediate causes of disease. For bodily illness caused by sin or rooted in the soul rather than the body, religious methods of healing through the mediation of holy persons or relics were the only effective and moral cure.

Where medical theory posited sickness as an imbalance of the natural state which should be corrected, Christian anthropology regarded *infirmitas* in all its forms as the normal and unavoidable human condition after the Fall.[77] In a frequently cited passage among medicine's critics, Pope Gregory the Great (ca. 540–604) had written: "the health

72 On the variety of criticisms directed toward scholars in Paris, see Baldwin 1970, and Ferruolo 1985. On criticism of medicine in particular, see Crisciani 1983.

73 See Amundsen 1996: 6–7. Discussing the Church Fathers on medicine, Amundsen argues that they did not say that medicine was sinful or false *per se*, but only that the soul should have priority over the body.

74 Crisciani 1983: 37–39.

75 For consideration of Greek medicine in early Christianity, see Temkin 1991, and Amundsen 1996.

76 Wallis 1995:117–124.

77 Agrimi and Crisciani 1978 and 1998 discuss the various religious and medical meanings of *infirmitas* and its relation to sickness.

of the soul may lie in the suffering of the body" (*salus cordis sit molestia corporalis*).[78] Disease is a cure for the soul, while perfect health might as well be an invitation to sin. For the Cistercian Bernard of Clairvaux (1090–1153), whose influence on church doctrine in the twelfth century can hardly be overstated, any concern for bodily health was in itself controversial.[79] From his point of view, excessive interest in the body constitutes immoral concern with, rather than simple use of, the material world. As in Gregory's remark, disease is actually "healthy" for the soul. The theoretical disapproval of Bernard of Clairvaux, Peter Comester, and others extended to practical subjects and particular therapies, such as frank discussions of sex (e.g., Constantine's *De Coitu*) and its potential place in a healthy regimen.[80] The difference between these moral priorities and those of the Chartrian naturalists is clear. For Adelard of Bath, the fact that the body drags down the soul is a motivation for seeking bodily health. Balance in the body brings clarity to the soul and the possibility of intellectual and moral improvement. In any case, ecclesiastical doctrine was quite clear that the health of the soul must always be placed before that of the body.[81]

Other uses of medical theory at the end of the twelfth century show that we should be careful not to exaggerate the attitudes of some vocal critics. Theological explanations of illness and mental phenomena were also capable of peacefully coexisting with the somatic etiology of the *medici*. Commentaries on the *Song of Songs* were a common location for use of medicine, where commentators often used medical language to explain the mental anguish of the lovers, even if they would never attempt a physical cure of such a spiritually uplifting passion.[82] If anything, the use of medical theory in Biblical exegesis shows that these commentators were confident enough in the subordinate position of medicine to not feel threatened by medical explanations. However, criticism helped set the parameters for the place of medical theory and practice in the schools in the following century. Within the schools, some scholars also contested the status of medical knowledge on intellectual grounds. Hugh of St. Victor (d. 1141), in his *Didascalicon* (1120s),[83] a highly influential review of all branches of knowledge, pointedly rebuts the claim of the *Isagoge*, denying that a theoretical part of medicine exists. He assigned medicine a place among the lowest mechanical arts.

---

**78** Quoted in Agrimi and Crisciani 1998: 174.

**79** Crisciani 1983: 36; Siraisi 1990: 14.

**80** Wack 1990: 58–59.

**81** For canons prohibiting physicians from offering medical advice that conflicted with Christian doctrine or the moral instruction of priests, see Jacquart and Thomasset 1985/1988: 196.

**82** Wack 1990: 21–23.

**83** Engl. transl. Taylor 1961: 78–79. On this classification, also see Weisheipl 1978: 473–474.

## Scholastic Medicine in the Early Thirteenth Century

### Changing Attitudes

Antagonism toward medicine subsided in the decades after 1200 for a variety of cultural and intellectual reasons. Broadly, there was a change in mentality with a newfound regard for materiality which resulted in profane sciences such as medicine and law being recast as licit and valuable activities, especially in the thought of the mendicant friars.[84] This change was part of a larger project of the renovation and reform of the church envisioned by Pope Innocent III (1198–1216) and his advisors. As in many of the other reforms at this time, the perceived threat from dualist "Cathars" or "Albigensians" was influential in the shift of ecclesiastical opinion.[85] To counter their dualist heresy that was understood to regard the material world – including the material aspect of human nature – with contempt, many of the reforms codified at Lateran IV in 1215 show a recommitment to the spiritual value of all physical things. This commitment included clarifications of core Catholic beliefs, such as God's creation of all things "both spiritual and corporeal" and Christ's physical presence in eucharist.[86]

Clerical rhetoric after 1200 began to stress the positive aspects of medicine. Sickness and human frailty were inevitable consequences of original sin, passed on physically by "vitiated seed", but medical knowledge is a divine gift to aid mankind in this fallen state and is essential in the pastoral obligation to care for the poor and infirm.[87] The Bible offered numerous examples of bodily healing, most obviously through the miracles of Jesus, and exhorted people to "honor the physician because he was instituted by God" (Eccl. 38.1). The care of the body was sometimes even regarded as a moral imperative; to do otherwise is to disregard God's gift of life.[88] Certainly this was not a novel idea; rather, the intensity of the previous hostility towards learned medicine was more exceptional. This pastoral obligation to care for the sick is exemplified in monastic infirmaries; at the same time many medical practitioners were also monks or priests, and their medical skill was justified as a divine gift to aid mankind.[89]

The same medical knowledge that enabled this religiously motivated therapy came to be valued by clergy for its ability to identify individual dispositions and tendencies.[90] In the confessional, priests used this information to understand (if not excuse) the causes of particular sins, and thus to assess individual culpability. Medicine supplied preachers with an entire category of commonplaces for their sermons where

84 Crisciani 1983: 35; French and Cunningham 1996.
85 French and Cunningham 1996.
86 Macy, 1994.
87 De Rijk 1967; Agrimi and Cristiani 1978 and 1998.
88 Temkin 1949.
89 Agrimi and Crisciani 1998: 182–192.
90 Agrimi and Crisciani 1978.

invocations of the care of the body are ubiquitous as an analogy for the care of the soul. Medical analogy reflected the idea that the learned practitioner was the holder of sacred knowledge about the body and has a position analogous to that of the priest, who cares for the soul. This shift in mentality occurred as the first universities were forming and helped with the establishment of medical faculties, who in turn were willing to submit to strict moral control, or at least to give sufficient lip service to the idea.[91]

## The First Decades of Medicine in the Universities

As the first universities were formed early in the thirteenth century, the origins of medical teaching in the schools were informed by these more positive religious attitudes toward medicine. After the decline of the school of Salerno toward the end of the twelfth century, Paris, Montpellier, and Bologna emerged as the most prominent medical schools in the thirteenth century and beyond, although there were several smaller schools elsewhere, including Padua.[92] At Paris and Montpellier, there is evidence for medical teaching by the middle of the twelfth century before they were formally organized into universities early in the thirteenth century.[93] At Bologna, the first evidence of medical instruction is a few decades later, with the medical faculty formally organized into a corporation by the 1260s.[94] At least at Montpellier and Paris, the administrators who composed the earliest statutes described the mission of the medical faculties in terms of their religious utility.[95] From the beginning, the medical faculty, like the other faculties, was subject to a fair amount of formal ecclesiastical oversight and admonition from moralists who constantly reminded scholars of the proper place of medicine in the Christian hierarchy of knowledge.[96]

Reorientation in the taxonomy of the arts and sciences along Aristotelian lines also contributed to medical theory being perceived as less of a threat to Christian doctrine. Theoretical knowledge (or science) in the strictest sense can only be about the eternal and unchangeable – the more unchangeable, the higher in the hierarchy of disciplines.[97] Disciplines that deal with the realm of the contingent or mutable, such as medicine, are lower on the scale of knowledge than theology, metaphysics, or natural philosophy. Aristotelian methodological doctrine further stipulated that an infe-

**91** Crisciani 1983: 36.
**92** For an overview of medicine in the universities, see Siraisi 1992, and Jacquart 1993/1998.
**93** On Paris, see Bullough 1957, and O'Boyle 1998. On Montpellier, see Bullough 1956, Demaitre 1975, and McVaugh 1990.
**94** On Bologna, see Siraisi 1981.
**95** Drayton 2001: 34–35.
**96** O'Boyle 1992.
**97** Lohr 2010.

rior science must accept the conclusions of a superior to which it was subalternated, as medicine was to natural philosophy.[98] Aristotle repeatedly uses medicine as an example of a practical art, in which it is only important to have contingent knowledge about what to do in a particular event rather than scientific understanding of universal principles. He frequently notes that a physician does not cure "man", but "Socrates".[99] Medical textbooks that came to be used most heavily from the first half of the thirteenth century did little to challenge this hierarchy of the disciplines and probably even encouraged it. One of the texts that would become most popular, Avicenna's *Canon*, argued that practical medicine is grounded in (Aristotelian) philosophy, but this relationship is not reciprocal. Unusual medical cases should not alter philosophical theories.[100] The physician was to "be familiar with natural philosophy, but not attempt to expound it", at least not in his role as a *medicus*.[101] Thus, for example, a physician should follow Galen's views about the brain when practicing medicine, but should accept Aristotle's view on the primacy of the heart as a higher truth.[102] At the same time, philosophers and theologians increasingly abandoned the use of medical sources or ideas, judging these to be irrelevant to their own work, with Albertus Magnus and Roger Bacon being important exceptions.[103] Most scholars, before the final decades of the thirteenth century, accepted the notion that philosophy is concerned with the truth, while medicine merely operates on the practical level of *utilitas*.[104]

Meanwhile, despite continued translation activity throughout the twelfth and thirteenth centuries, the contents of medical curricula did not stray far from the traditional *Articella* for several decades. After Constantine, notable translators included Gerard of Cremona (ca. 1114–1187), who translated from Arabic in Toledo, and Burgundio of Pisa (1110–1193), translating from Greek.[105] Translations included Avicenna's *Canon* and the *Liber ad Almansor* by Abū Bakr al-Rāzī (Latin: Rhazes or Rasis, 865–925), works which were not immediately popular, but would eventually replace the *Pantegni* as the most popular medical encyclopaedias over the course of the century.[106] Gerard and Burgundio also translated many Galenic treatises and Arabic adaptations of Galenic treatises,

---

**98** Livesey 1994. For Aristotle's division of the arts and sciences as interpreted in the Middle Ages, see Weisheipl 1978: 464–468.

**99** For example, Aristotle, *Metaphysica* 1 = 981a17–21 (ed. Tredennick 1933: 5). A similar point is made in the *Ethica Nicomachea* I.6 = 1097a12–14 and V.9 = 1137a13–17 (ed. and transl. Rackham 1926: 25 and 311).

**100** See Daston and Park 1998 for a discussion of how strange and particular medical facts were excluded from philosophy.

**101** Siraisi 1981: 154, citing the *Canon* 1.1.1.2.

**102** On the dispute about the heart and brain, one of the most significant differences between Aristotelian and Galenic physiology, see Jacquart 2003.

**103** Jordan 1992.

**104** Garcia-Ballester 1995: 141–148.

**105** On Gerard, see Burnett 2016a. On Burgundio, see Buellens 2016.

**106** Avicenna, *Liber canonis* 1507/1964. Rhazes, *Almansor* 1544/1973. For a history of Avicenna's *Canon* in the West, see Siraisi 1987. The *Canon* was not widely used until closer to the middle of the thirteenth

including the *De complexionibus*, *De interioribus* and other texts which would be extremely important for understanding Galen's theories of human nature. Like the *Canon* and the *Almansor*, these Galenic texts were slow to be incorporated into the medical curriculum. Before the end of the thirteenth century, most learned physicians probably only read the *Tegni* (part of the *Articella*) and later, Galen's commentaries on the Hippocratic *Aphorisms* and the *Prognostics* (which were appended to the *Articella* during the course of the thirteenth century, in what is known as the *ars commentata*).[107] In the statutes of 1270–1274 at Paris, the required texts in the medical curriculum had only been supplemented by a few practical texts (e.g., on diets and regimen) and the situation was similar elsewhere.[108] As Michael McVaugh observes, the availability of the new texts does not necessarily entail their actual assimilation. Instead, most of the newly translated Galenic treatises were rarely used before the end of the thirteenth century, when a core of essential texts would finally emerge, and the recovery of Galen's authentic views became a priority.[109]

In contrast to the widespread interaction between medicine and philosophy in the twelfth century, medical interests at the universities seem to have remained almost exclusively practical for the first half of the thirteenth century.[110] No commentaries on Galen's *Tegni*, the most philosophically rich text of the *Articella*, survive from Paris in this period.[111] It seems that many physicians were primarily interested in professional careers.[112] Most of the written work that survives from before 1250 or so consists of case studies, *consilia*, and recipes arranged for the teaching of practical medicine.[113] In a reminder of the religious oversight of the medical faculty, medical writing from this period is often careful to acknowledge the dependance of medical efficacy on divine grace and to insist on a religious component to the task of the *medicus*. Peter of Spain (fl. mid 1200s and probably *not* the person who became Pope John XXI in 1276, or the famous logician) reminds his readers that because Christ himself is the highest *medicus*, it is not only unprofessional to be unlearned, but sinful: incomplete knowledge may impugn God.[114]

century or later in Paris, Montpellier, and Bologna (Siraisi 1987: 50–53). Jacquart and Micheau 1990: 11–12.

**107** O'Boyle 1998: 103–123 traces these additions.

**108** O'Boyle 1998: 20–24.

**109** McVaugh 2019.

**110** Jacquart 1993/1998: 215. A substantial minority of medical students went on to study canon law or theology according to O'Boyle 1998: 77.

**111** Ottosson 1984: 32–34.

**112** Jacquart 1993/1998: 210–214.

**113** O'Boyle 1998: 78. On *consilia* see Agrimi and Crisciani 1994.

**114** Peter of Spain, *Thesaurus Pauperum*, ed. Pereira 1973: 79.

### Aristotelian Natural Philosophy

Original work in natural philosophy shifted to the study of Aristotle and Arabic philosophy inspired by Aristotle. This story of the recovery of Aristotle's philosophy and the decades of accompanying controversy, especially in the arts faculty at Paris, is well known in the intellectual history of this period.[115] As we saw, some of Aristotle's logic and a few of his physical works had been used in Salerno by 1150. Further translation of Aristotle's philosophy continued and, by the middle of the century, almost all of Aristotle, including several treatises attributed to Aristotle and Arabic interpretations or continuations of Aristotelian thought, had been translated and studied in Paris.[116] Aristotle's texts supplied the conceptual instruments for a far more complete understanding of the natural world than the medical theory of the previous century, and the structure of the Aristotelian corpus (or at least as this structure was interpreted) came to provide the organization for all of philosophy. According to taxonomies of knowledge from early in the reception of Aristotle, philosophy was divided into rational, natural, and moral philosophy, each with their own sub-branches.[117] Rational philosophy contained the subjects of the *trivium*, which included the logical texts of Aristotle's *Organon*. Natural philosophy contained metaphysics, mathematics, and physics, which included works on eternal celestial motion, change in the sublunar world, and a whole range of biological works, which would be most important for medicine. Moral philosophy included Aristotle's *Ethica Nicomachea* which, as a text about the practical life of the soul, would also attract medical interest.[118]

These texts opened a new era in the philosophical treatment of human nature and the connection between body and soul. Of particular importance for this topic, the *De anima* provides an overview of all soul, not just human, divided into the vegetative, sensitive, and rational souls and their associated faculties.[119] Avicenna's own *De anima*, which was widely seen as a commentary on Aristotle, was equally influential and heavily influenced interpretation of this.[120] The treatises of the *Parva naturalia* (Short Treatises on Natural Sciences) deal with a range of psychological and

115 For an overview, see Van Steenberghen 1955 and 1991.

116 For an overview of the translation of Aristotle, see Dod 1982, and Pasnau 2012. A list of Aristotelian works (including pseudonymous works) translated into Latin can be found in Pasnau and Van Dyke 2010: 2.793–797.

117 Lohr 2003: 16–17.

118 On interest in the *libri morales* among physicians at Bologna, see Siraisi 1976. According to Siraisi 1976: 106, Taddeo Alderotti translated a Latin compendium of the *Ethica* into Italian.

119 On the introduction and use of the *De anima* in the West, see De Boer 2013. It was translated into Latin three times in this period: from the Greek ca. 1150 by James of Venice (fl. 1125–1150), from the Arabic along with Averroes' *Long Commentary* ca. 1230 by an unknown translator, and again from the Greek ca. 1266 by William of Moerbeke (d. ca. 1286) (De Boer 2013: 16).

120 Avicenna. *Liber de anima*. ed. van Riet 1968 and 1972. On the Latin reception of Avicenna's *De anima*, see Hasse 2000.

physiological topics common to the soul and body, including sense and sensible things, memory and recollection, sleep and waking, dreams, prophecy in sleep, length and shortness of life, youth and old age, life and death, and respiration and inspiration.[121] These texts were usually understood as amplifications of particular parts of the framework of the *De anima*. Finally, the texts collectively known as the *De animalibus* in nineteen books, including the *Historia animalium* (Books 1–10), the *De partibus animalium* (Books 11–14), and the *De generatione animalium* (Books 15–19) provided Aristotle's most systematic anatomical and physiological material (*De partibus*), as well as information particular to various animals (*Historia*) and reproduction (*De generatione*).[122] The combined force of these texts suggest that the science of the soul should be situated in animated bodies and their biology, although many medieval (and modern) philosophers were more interested in metaphysical questions concerning the separability and immortality of the soul.

The use of Peripatetic natural philosophy quickly provoked censure from ecclesiastic authorities with authority over the university at Paris, where the new texts first came into widespread use. In 1210 and again in 1215 scholars were forbidden from teaching Aristotle's natural philosophy or the commentators of those works.[123] As part of the 1210 censure, the *Quaternuli* of David of Dinant (ca. 1160–ca. 1217), which made use of the works in question in its naturalistic approach to the body-soul relationship and natural analysis of miracles, were ordered to be burned.[124] In 1228 and again in 1231, Pope Gregory IX (1227–1241) sent a letter to the regent masters in theology at Paris warning them not to allow this philosophy to corrupt the truth of scripture, while ordering students in the lower arts faculty not to read the *Libri naturales* at all.[125] He assigned a commission to purge the texts of error, but this work appears to

---

**121** Aristotle, *Parva naturalia*, ed. and Engl. transl. Beare and Ross 1908. Most of the *Parva naturalia* was translated from the Greek by James of Venice in the mid-twelfth century, and the entire collection was translated again from Greek by William of Moerbeke in the 1260s. See De Leemans and Klemm 2007: 162–163.

**122** These texts were translated from the Arabic before 1220 by Michael Scot (d. ca. 1232) and from Greek by William of Moerbeke in the 1260s. See Buellens 2007: 136–137. Most of Scot's translations of the *De animalibus* have been edited: *Historia animalibus*, Books I–III, 2020; *De partibus animalium*, 1998; and *De generatione animalium* 1992, all ed. van Oppenraay. Books IV–X of the *Historia* are in preparation. Moerbeke's translation of the *Historia animalium* is available in two volumes of the *Aristoteles Latinus*, ed. Buellens and Bossier 2000 and 2021. For the Greek and English Loeb volumes, see *History of Animals*, ed. and transl. Peck and Balme 1965 and 1991; *Parts of Animals*, ed. and transl. Peck 1937; and *Generation of Animals*, ed. and transl. Peck 1942.

**123** Van Steenberghen 1970: 66–77. The texts of these censures are reprinted in *Chartularium Universitatis Parisiensis* I, ed. Denifle and Chatelain 1889: 70 and 78–79. The ban only applied to natural philosophy: in 1215, the Old and New Logic, the first three books of the *Ethica*, and part of the *Topica* were required reading.

**124** On what little is known about Dinant and the *Quaternuli*, see Maccagnolo 1988.

**125** Van Steenberghen 1970: 78–88. For these letters, see *Chartularium Universitatis Parisiensis* I, ed. Denifle and Chatelain 1889: 114–116 and 143–144.

have never been completed, probably due to the impossibility of the task and the continual arrival of new Aristotelian texts and commentaries. It may also be that continuing concern about heresy and the need to confront them with rational arguments reduced the motivation to censure pagan philosophy.

By the 1240s, these censures were no longer in force in Paris when Roger Bacon (ca. 1220–1292) lectured with impunity on Aristotle's works, and knowledge of Aristotle was rapidly becoming the norm for any master aspiring to the higher faculties of theology, medicine, or law.[126] In the next two decades, Bacon and the Dominicans Albertus Magnus and Thomas Aquinas (1225–1274) were only the most prominent among a host of scholars to integrate the Aristotelian corpus into the mainstream of philosophy and theology. In 1255 the arts faculty restructured its curriculum to include in its required lecture list all the known books of Aristotle, including his natural philosophy.[127] The arts school at Paris had effectively transformed itself into a philosophy school based largely on the thought of Aristotle and his interpreters.

The medical faculty in Paris also adopted relevant texts, although with much less controversy, and the *De animalibus* and Book 4 of the *Meteorologica* joined the *Articella* as required reading in 1274.[128] As important as these texts were for medicine – in particular, for understanding differences between Aristotelian and Galenic biology – they were relatively neglected by philosophers in the arts faculty, which had a much closer relationship to theology than to medicine, so that Aristotle was often "theologized" and abstract questions were privileged.[129] The works on animals, for example, have a relatively scant commentary tradition compared to other Aristotelian texts, and seem to depend on how relevant any particular scholar considered them to be for the study of the human soul.[130] However, there were some extremely influential exceptions by Peter of Spain and, most popularly, by Albertus Magnus.[131] These authors would all be influential in the medical philosophy of the following century.

The biological texts of Aristotle became standard reading in medical instruction throughout the rest of Europe as well. At Montpellier, the necessity of the study of Aristotle in conjunction with medicine was made clear in statutes from 1240 and again 1309.[132] At Bologna, Aristotelian philosophy and the scholastic methods that had developed at Paris arrived later, but would establish a closer relationship with medi-

**126** On Bacon's lectures, see Wippel 1977: 172.

**127** See Denifle and Chatelain 1889: 278.

**128** O'Boyle 1998: 24.

**129** French and Cunningham 1996.

**130** De Leemans and Klemm 2007: 164–166.

**131** On Peter of Spain's commentaries on the *De animalibus*, see De Asua 1999. For an edition, see Navarro Sanchez 2015. On the identity of Peter of Spain, see D'Ors 2001. For Albertus' commentary, see *De animalibus*, ed. Stadler 1916 and 1920 (Engl. transl. Kitchell and Resnick 2018). For Albertus' disputed questions on animals, see *Quaestiones super de animalibus*, ed. Filthaut 1955 (Engl. transl. Kitchell and Resnick 2008).

**132** Siraisi 1990: 66.

cine than at Paris.[133] By the 1260s, philosophy was still not very well established at the public *studium*, when Taddeo Alderotti (ca. 1210–1295), the most prominent master of the generation before Pietro, helped institute the study of Aristotle's *libri naturales* in conjunction with the Salernitan program of medical study and the "new" Galen and Avicenna.[134] Thus, at Bologna, medicine and philosophy became part of the same faculty and shared a common origin.

Although there was little controversial work done in medical philosophy in the first several decades of medicine in the universities, the institutional stability they provided laid the groundwork for the creative period that would follow. By 1270, the traditional canon of medical texts from the twelfth century had been fully digested and the newer texts of Galen and Avicenna, coupled with the integration of Aristotelian philosophy into medical teaching, provided the basis for new directions. Medicine was also influenced by the scholastic methods inspired by Aristotelian logic. The techniques of *lectio, expositio, questio,* and *disputatio* would be employed in the interpretation of the new medical texts and are increasingly visible towards the end of the century. These methods supplied the tools for the analysis of differences in the growing list of authorities, a central theme of the decades around 1300.

## New Ambitions, 1270–1320

The late thirteenth and early fourteenth centuries were decisive in the formation of philosophical medicine, as they were for all areas of scholastic medicine. There was a marked increase in the sheer quantity of writing that survives from these decades, and several celebrated *medici*, led by Taddeo Alderotti and his students in Bologna and Arnau of Vilanova at Montpellier, produced commentaries that attempted to synthesize the ever-expanding collection of philosophically informed medical texts.[135] The assimilation of what has come to be called the "new Galen" was the most significant project of these years.[136] This was a difficult process. Galen was famous for his rambling style, filled with anecdotes and polemics. The task of distilling his thought into a neat outline of doctrines separated into theory and practice (a distinction that is not present in Galen's own work) had been one of the main accomplishments first at Alexandria and later in the Arabic encyclopedias.[137] It is no surprise that the *Isagoge* and *Pantegni*, and then the *Canon*, had been more influential in introducing Ga-

**133** Siraisi 1981: 7–9.
**134** Siraisi 1981: 96–117.
**135** For overviews of this development, see Agrimi and Crisciani 1988: 11–20, and Jacquart 1993/1998. For Bologna, see Siraisi 1981. For Montpellier, McVaugh 1990.
**136** On the "new Galen" and its assimilation, see Garcia-Ballester 1998, Nutton with Bos 2015: 91–100, and McVaugh 2019.
**137** Temkin 1973: 67–94.

lenism into Latin scholarship than Galen's own work. Of Galen himself, physicians would have only known the relatively short *Tegni* and his commentaries on the Hippocratic *Prognostics* and *Aphorisms*. Starting in the 1220s, Avicenna's huge *Canon* began to increase in popularity as the best guide to Galenic thought and eased the path for the real thing.[138] By the 1260s in Montpellier, Galen's texts were beginning to be cited regularly. Sometime before ca. 1285 in Paris, Jean of St. Amand (ca. 1230–1303) produced summaries of many of Galen's treatises and a large glossary of medical terms with citations of their use, indicating that knowledge of Galen was now coming to be expected of medical students.[139] By the end of the century, Taddeo and Arnau produced detailed commentaries on several of Galen's treatises, more completely assimilating them into medical instruction, as a group of essential Galenic texts emerged.[140]

## Medicine as Science

In Galen, scholastic physicians possessed a figure equal in authority to Aristotle, and in Galen's texts, they found a connection between medicine and philosophy that was quite unlike the merely technical status assigned to medicine by their earlier contemporaries. It had long been a convention to address the nature of medicine at the beginning of medical texts, as least cursorily. However, during this period, debate about whether medicine is more theoretical or practical, more like a science or more like an art, became prominent in the prefaces to encyclopedias and commentaries on certain texts, especially the *Tegni.*[141] In Bologna, Taddeo Alderotti, in the introduction to his commentary on the *Tegni*, confidently claimed that medicine is a *scientia* in the Aristotelian sense, reaffirming the importance of the theoretical part of medicine.[142] Medicine, Taddeo wrote, is strictly demonstrable knowledge of the human body.

Pietro d'Abano makes a similar but longer and more passionate case for the high status of medicine, arguing that not only did it meet the Aristotelian standards of a science, but also that it was the most excellent of the arts.[143] He devotes the first ten questions (*differentiae*) of the *Conciliator* to the analysis of medicine as a discipline, more than fifteen double-sided folia in the Renaissance printed editions. After con-

**138** Siraisi 1987: 43–45.

**139** On the evidence for the integration of Galen into medical texts, see McVaugh 2019. For John's work, see John of St. Amand's *Concordantiae*, ed. Pagel 1894. For the *Abbreviationes librorum Galieni*, in John's *Revocativum memoriae*, see McVaugh 1990: 63–67.

**140** Traditionally, there were sixteen essential texts by Galen (Temkin 1973: 69). But see Nutton 2015: 91–100, on the variability in the transmission of texts ca. 1300.

**141** On the range of opinions on the nature of medicine, see Ottosson 1984: 65–126, and Agrimi and Crisciani 1988: 137–155.

**142** Taddeo's full position is discussed in Siraisi 1981: 118–125.

**143** Pietro's analysis of the methodology and status of medicine has attracted more attention than most aspects of his thought. See Randall 1940/1966, Wallace 1995, Ottaviani 2004, and Bottin 2013.

cluding in the first two *differentiae* that it is necessary for the physician to be well-versed in other speculative sciences and logic, *differentia* 3 asks, "Whether medicine is a science" (*An medicina sit scientia*). Here, he attempts to reconcile the teaching of Aristotle's *Analytica posteriora* with Galen's *Tegni* in his own understanding of medicine as science. He argues that medicine should be understood as science in two main ways: as a body of knowledge that establishes a habit in the intellective soul (*habitus animae intellectivae*) and in the sense that it has the appropriate methods and objects. Pietro's argument is extremely complex and, in combining Galen and Aristotle, he almost certainly does not represent either authority accurately.[144] He defends a method of discovering scientific knowledge that strays from Aristotle's definition of absolutely certain knowledge and instead has lower standards of being based on what is true most of the time (*ut plurimum* or *possible in pluribus*).[145] At the same time, he equates Galen's resolutive and compositive methods from the *Tegni*, which are usually interpreted as methods of demonstration in teaching, with Aristotle's methods of *quia* and *propter quid*, which are more properly methods of scientific discovery.[146]

However convinced we are by Pietro's reading of Galen's methods, the rhetorical force of his claims for medical authority is notable. With characteristic invective, he says that those people who deny that it is possible for medicine to be a science (he is referring here to knowledge about internal processes in particular) are "dense" (*grossi*) and have "weak intellects" (*hebetis intellectus*).[147] Just because unintelligent people are not able to understand the scientific nature of medicine does not mean they should judge that others cannot. In fact, the subject of medicine – the human body – is the most noble object of knowledge in the world and is the most attuned to the human intellect. Taking a shot at theologians, he contrasts the subject of medicine, the human body, the most noble body in the world, with things about God and the intelligences, which exceed the ability of our intellect.[148]

Certainly not everyone agreed with Taddeo and Pietro. Many of Taddeo's own students, who later went on to become eminent physicians themselves, regarded the claim that medicine is a science as not completely valid. Bartolomeo da Varignana (d. after 1321) and Turisanus (or Pietro Torrigiano, d. ca. 1320) each questioned the identi-

---

**144** As we will see in the following chapters, this will not be the last time that Pietro makes a rather strained argument to claim that Galen and Aristotle agree.

**145** See Ottaviani 2004: 15, and Bottin 2013: 319–325.

**146** On the identification of Galen's resolutive and compositive doctrines in the *Tegni* with Aristotle's *Analytica posteriora*, see Ottosson 1984: 98–124.

**147** *Conciliator*, diff. 3 (f. 7r, col. 1, B). Also see diff. 5 (f. 9r, col. 2, A), where he notes that "some people dishonor this science especially . . . it is vituperated by idiots . . . who are mixed up because their senses are unskilled, and their intellects are blocked".

**148** *Conciliator*, diff. 3 (f. 7r, col. 1, D): "Est enim illic subiectum, quod omnium est, quae in mundo nobilissimum corpus, puta humanum sanabile (differentia 6). Capitur etiam ab intellectu, cum non lateat nos propter eius debilem entitatem, velut materia prima, aut quia potentiam superat intellectus, ut ea quae sunt Dei et intelligentiarum, sed sit quid medium intellectui conterminum et propinquum."

fication of medicine with natural philosophy and emphasized the ultimate aim of medicine to maintain and restore health, rather than produce knowledge for its own sake.[149] Arnau of Vilanova argues that Galen himself warned against the overuse of theory in comparison to the facts of experience.[150] Pietro and Taddeo were reacting against the low opinion of medical philosophy formed in previous generation. In doing this, they stressed the theoretical side of medicine more than physicians in the following generation would accept.[151]

## Galenism and Aristotelianism

This period was also marked by the recognition of differences between the Galenic and Aristotelian traditions. The explanation of these differences was closely tied to arguments for the elevated status of medicine, because disagreement with Aristotle had the potential to diminish Galen's authority. Differences included the number of principal organs, reproductive theory, the primacy of the heart or brain, the anatomy of the nervous system, the physiology of sensation, among many other particular anatomical and physiological doctrines.[152] In scholasticism, where the ideal mode of treating discrepancy was harmonization, the extent of some of these differences constituted a significant challenge.

Avicenna's *Canon* and his *De animalibus* (trans. ca. 1220 by Michael Scot [d. ca. 1232]), built on earlier Arabic writers to provide the most complete description of differences, as well as a model for synthesizing the two traditions.[153] In Latin, the first comprehensive attempt to deal with disagreements seems to have been the physician, Peter of Spain, in the questions of his commentary on Aristotle's *De animalibus*, probably composed 1246–1249 in Sienna.[154] More influentially, Albertus Magnus' own *De animalibus*, composed between 1257 and 1268, and his *Quaestiones super de animalibus*, which derived from his teaching in Cologne in 1258 and may have drawn on Peter of Spain's questions, further publicized the areas of conflict.[155] By the end of the thirteenth century, Averroes' (1125–1198) encyclopedic *Colliget* (trans. 1284) supplied a

**149** Siraisi 1981: 126–137.

**150** McVaugh 1990: 69–75.

**151** As Ottosson 1984: 124–126 comments.

**152** On the major differences, see Siraisi 1981: 186–201.

**153** On the transmission of *De animalibus*, see Van Oppenraay 2017. The text has been printed twice in Latin, in 1500 and 1508. On Michael Scot, see Burnett 2016b.

**154** De Asúa 1999. According to De Asúa, at least 26 questions in this commentary mention conflicts between Aristotelian and Galenic doctrines.

**155** Albertus Magnus, *De animalibus*, ed. Stadler 1916 and 1920 (Engl. transl. Kitchell and Resnick 2018). *Quaestiones super de animalibus*, ed. Filthaut 1955 (Engl. transl. Kitchell and Resnick 2008). On the dating, see Kitchell and Resnick 2018: 35. On Albertus' treatment of specific doctrines, see De Asúa 2013.

long list of particular disagreements, usually criticising Galen for his lack of philosophical rigor.[156] These differences were opportunities to reflect on the relationship between medicine and philosophy.

Avicenna suggested that medicine and philosophy have different perspectives, with medicine seeing the body as it is now, and philosophy considering it on a more absolute and abstract level saying, "Although their [the physicians'] view appears to be superficially the more plausible, the philosopher's [Aristotle's] is really the correct one."[157] Albertus essentially agreed, saying of the different sciences, "If the discussion concerns medicine, I would rather believe Galen or Hippocrates, and if it concerns things of nature, Aristotle or anyone else experienced in natural things."[158] For matters of faith, Augustine should be preferred.[159] Taddeo, who argued more forcefully than most for the scientific status of medicine, also argued that it was important to realize that usually Aristotle was more correct.[160] Arnau of Vilanova, guided by Avicenna, argued that it did not matter to the physician whether Aristotle or Galen is more correct in an absolute sense, because, "The physician orients his understanding and theorizing toward practice, and hence is not concerned to understand more theory than is sufficient to govern or correct his practice . . . ."[161]

Pietro's *Conciliator* is almost certainly the apex of this conversation around 1300. In comparison to his contemporaries, he more often finds the opinions of Galen and Aristotle to be directly comparable and uses a variety of methods to resolve apparent disagreements, as we will see in the following chapters. He sometimes tries to show that Galen and Aristotle are using different terms to talk about the same thing (as in his thoughts on methodology above), or that they do not actually disagree (on complexion, Chapter 2), or that medicine supplies a material account, rather than a contradictory doctrine (on spiritus and faculties of the soul, Chapters 3 and 4). Other times, he seems to distort either Galenic and Aristotelian thought (or both).

## Philosophy at Paris

Philosophical interests outside of the medical faculty in Paris also had a large impact on the thought of several learned physicians.[162] Certainly the intellectual milieu at Paris in the final decades of the thirteenth century contributed to the particular blend of philosophy and medicine in Pietro's thought. Earlier concerns about heterodoxy at

**156** Bürgel 1967: 291–297. Averroes, *Colliget* 1562, repr. 1962.
**157** Avicenna, *Canon* I, 1, 5, 1 (quoted in Jacquart 1993/1998: 218).
**158** Albertus, *Super II sententiae* 13, C, 2 (quoted in De Asúa 2013: 296).
**159** Siraisi 1981: 191.
**160** Siraisi 1981: 192.
**161** Arnau, *De Intentione medicorum* 2.1 (quoted in McVaugh 1990: 78).
**162** Siraisi 1981: 154–186.

Paris had resurfaced in the 1270s, culminating in the condemnation in 1277 of 219 propositions allegedly being taught in the arts faculty, perhaps the most famous single event in the intellectual history of this century.[163] The serious nature of the conflict is evidenced by reactions to it within the university. Earlier, in 1272, the arts faculty had tried to prevent trouble by forbidding its members to dispute "any purely theological question", and also stipulated that any question "which seems to touch on both faith and philosophy" must be answered according to faith.[164] By 1277 the antagonism toward the "radical" arts masters seems to have branched out to include less-radical philosophers and theologians who made extensive use of Aristotelian ideas.[165]

The censured propositions cover a wide range of material, but in general they are considered to be an attempt to curb rationalist inquiry pursued without regard for Christian doctrine. Many of the condemned items extol the pursuit of philosophy, e.g., "That there is no more excellent state than to be at leisure in philosophy."[166] Some circumscribe God's omnipotence with natural causes or assert that God cannot alter the universe as it is now ordained. Other censured articles include the Aristotelian proposition of the eternity of the world, the Averroist idea of the unity of the intellect, several doctrines which seem to reduce the freedom of the will, attribute too much causal power to celestial bodies, or directly oppose Christian morality. Modern scholars have long debated the effect of the condemnation on the thought of the period that followed.[167] Whatever the long-term effects, it is clear that, in the generation after 1277, the atmosphere at Paris remained highly contentious and that many of the censured ideas continued to be in circulation. Pietro himself reports that many "depraved" people continued to hold Averroes' interpretation of the intellect yet pretended otherwise in accord with the condemnation (*cum lege*).[168]

It requires little effort to find instances of the censured ideas reflected in Pietro's work. When considering the influence of the arts faculty on his thought, it is notable that the condemned ideas do not seem to have been taught within the medical faculty, or at least the medical faculty was not a target of condemnation; very few of the condemned items bear any relationship to medical doctrine at all.[169] In the *Conciliator*,

163 For the complete contents, cross-referenced with earlier lists, see Hissette 1977. For a summary of the condemned ideas, see Wippel 1977, and Thijssen 2023.

164 *Chartularium Universitatis Parisiensis*, ed. Denifle and Chatelain 1889: 1.499.

165 For discussion of possible targets outside the arts faculty, see Hissette 1977, Van Steenberghen 1980, Wippel 1995, and Normore 1995.

166 "Quod non est excellentior status quam vacare philosophie." (Item 1 in Hissette 1977: 15).

167 For a summary of interpretations, see Thijssen 2023. Some important arguments for the effect of the condemnation include Grant 1979, Murdoch 1991, Thijssen 1998, and Bianchi 1998 and 2003.

168 *Conciliator*, diff. 3 (f. 8r, col. 1, B): "Dicendum quod id pravae ponentes unitatem intellectus non sic angit, sed unitatem plualitatem ipsius profitientes cum lege."

169 Cadden 2001: 72–73, and Jacquart 2001 suggest a few specific items that are connected to medical issues. Even these items were not directed at the medical faculty, but the arts.

Pietro includes an analysis of miracles (walking on water),[170] a qualitative explanation of transubstantiation,[171] a natural theory of moral corruption of the human species over time, with derogatory comments about the theological approach to this same subject,[172] and a strong defense of celestial causes, that could easily be taken for astrological determinism, with more polemic against theologians.[173] His *Expositio Problematum* includes a remarkably naturalistic discussion of sexual desire, and same-sex desire in particular.[174] At least some aspect of each of these lines of inquiry, probably formulated by Pietro in the 1290s, had been explicitly condemned in 1277. All of this suggests that, while at Paris, Pietro was influenced by current ideas outside of the medical school, but that he seems to have been surprisingly indifferent to the threat of condemnation.

Beyond the specific doctrines circulating at Paris, Pietro's framing of his project as whole was almost certainly influenced by the controversial philosophical ambitions to achieve a kind of natural perfection in this life. This notion was based on Aristotle's *Ethica Nichomachea* 10, in which he discusses the perfect operation of the intellect. Philosophers in the arts faculty used this text to defend an elitist ethical ideal: the *summum bonum* in life consists of dispassionate contemplation of abstract ideas in order to attain the scientific knowledge that would lead in turn to the perfection or "beatitude" of the soul.[175] Thus all affections of the body and soul, all contingencies that produce them, must be banished to attain happiness.[176] Arts masters reiterated this ideal in various ways, optimistic that the exercise of their own professional activities was identical with the pursuit of human perfection. Similar sentiments had been censured in 1277; however, far from dying out in its aftermath, the idea gained in popularity as a means for philosophers to assert their independence from theology.[177] For some intellectuals in the arts faculty, the objective of achieving *beatitudo* through the actualization of their natural rational powers became something of a job description.[178] It is a theme echoed repeatedly in the prologues to philosophical works published at this time.

---

**170** *Conciliator*, diff. 21.
**171** *Conciliator*, diff. 156.
**172** *Conciliator*, diff. 9.
**173** *Conciliator*, diff. 10.
**174** *Expositio*, Book 4. For a comprehensive discussion of this text, see Cadden 2013.
**175** See Kuksewicz 1986; and Celano 1986 and 1990, and Bianchi 1987.
**176** Bianchi 1987: 185–186.
**177** Aertsen, Emery, and Speer 2001: 10, call the notion of philosophical perfectibility to be one of the major themes of 1277: "There can be no doubt that philosophical doctrines promising 'divinisation' of the human intellect, perfection in the natural virtues, and the possibility of felicity in this life, all without consideration of the sinful human condition of benefit of grace, were a serious ethical challenge to the teachings of Catholic moral theology and eschatology."
**178** See especially Bianchi 1987.

Pietro invokes this idea of beatitude in the prologue to his *Conciliator*. Citing Aristotle's *Ethica* 10, he remarks that happiness or *beatitudo* is a state of perfection, in which all "goods" have been acquired; specifically, *beatitudo* is an operation of the soul following from perfect virtue and a result of our natural capacity for knowing.[179] While agreeing with his colleagues that everyone seeks this kind of happiness (*foelicitatem appetere omnes*), he counters that most people do not ever come close to this narrowly defined conception of happiness: it is simply not useful for real life:

> Each person seeks to be happy, especially [pursuing] the perfect [form of happiness]. And yet: their lives are lacking and naturally depraved; at the same time their foundations are out of proportion, caused by intrinsic and extrinsic forces, and their actions are completely deficient of virtue.[180]

Rather, most people are "slaves by nature, not only in their bodies, but also in their souls".[181] With this preface, Pietro offers his medical philosophy as an alternative and more realistic approach to correct these natural deficiencies and achieve happiness in this life. The remainder of the *Conciliator* and the rest of his work reflect this intent.

179 *Conciliator*, Prologue (f. 1r, col. 1, B–C) "Est denique foelicitas operatio quaedam animae et usus secundum virtutem perfectam." and "Beatitudo est status omnium bonorum aggregatione perfectus. Palim: cunctis hominibus natura insitum est desiderium ad sciendum, quod beatitudinem (ceu tactum) terminatur."

180 *Conciliator.*, Prologue (f. 1r, col. 1, C): "[F]oelicitari appetit unumquodque quod autem perfectum, preacipue Adhuc demum: vita privantur, quorum naturaliter prava, simulque improportionata extant principia ab intrinsecis et extrinsecis causata et actus virtutis deficientes omnifariam."

181 *Conciliator*, Prologue (f. 1r, col. 1, C–D): "Quidam namque non solum corpore, verum et anima natura fere sunt servi. Fortunam demum talium similiter impeditivam, ingenuitatem fore divitias, potentias ac his opposita omninoque fortunium et infortunium (*Rhetoricorum* testatur secundo). In qua, vim geneseos sydeream esse puto non immerito includendam. Nonnulli enim vita privantur ac beatitudine bona decepti aestimata fortuna velut et prava. Nam etsi ea foelicitati ministret praesidium, quo, quanto, quali, et ubi, et quando expedit modificare."

# Chapter 2
# Complexion between Body and Soul

In commenting on entry 30.4 in the *Expositio Problematum*, where the Aristotelian text states that the hands and intellect were given by God to humans as natural organs for attaining the virtues, Pietro interjects that complexion should be added to these two special features of humanity: "With these two instruments, [God] also gives a third, namely complexion, in virtue of which the foresaid two function."[1] Again in *differentia* 57 of the *Conciliator*, where Pietro includes a hierarchical list of all the faculties of the soul, he makes the idiosyncratic addition of the elements and their complexions to a list that is otherwise fairly conventional.[2]

As these two highly unusual interventions suggest, the most fundamental concept for Pietro's thinking about human nature is the medical notion of "complexion", "the effect resulting from alteration in the qualities of the elements in a mixture", that is, the sum of the hot, cold, wet, and dry qualities in a body.[3] *Complexio* (*temperamentum*, *mixis*, or κρᾶσις) would have been familiar from a variety of ancient and medieval philosophical texts, especially in discussions about the survival of individual elements in mixtures.[4] In medicine, complexion theory increased in prominence at the end of the thirteenth century, providing a way to organize all other parts of theory and practice.[5] All species of natural bodies and their organs were healthy so long as they possessed an appropriate balance of qualities; illnesses and therapies were organized according to the kind of effect they had on these complexions. For the educated doctor, complexion theory was a way to claim knowledge of internal processes which could not otherwise be known directly.[6] Textbooks assured physicians that the fingers are the most temperate part of the body in order to defend the reliability of touch to detect complexions.[7] As the concept increased in sophistication, the primary thera-

1 *Expositio*, 30.4: "Cum his etiam duobus instrumentis dedit et tertium, complexionem scilicet, in cuius quidem virtute operantur pretacta duo, et precipue manus. Unde Avicenna Primo Prime [*Canonis*]: Donavit deus homini nobiliorem complexionem que in hic mundo foret possibilis invenire."

2 *Conciliator*, diff. 57 (f. 84r, col. 1, B): "Ex his itaque diligenter computatis apparet, quod virium animae catena 34 aut 35, cum eisdem deservientibus constituitur annulis, quatour videlicit qualitatibus . . ." For a discussion of the entire "chain" of the faculties of the soul, see Chapter 4: 75–83.

3 *Conciliator*, diff. 17 (f. 26r, col. 2, B): "Dicitur etiam, quod complexio est effectus ex elementorum qualitatibus alteratione resultans in mistis."

4 The term "*complexio*" as a translation of "κρᾶσις" entered Latin medicine in the twelfth century and became the norm by the thirteenth. See Jacquart 1984/1997.

5 On the centrality of complexion in thirteenth- and fourteenth-century medicine, see McVaugh 1975, Ottosson 1984: 127–194, and French 2001: 89–141.

6 French 2001: 90, emphasizes the idea that such knowledge of internal processes gave scholastically trained physicians a professional advantage over other practitioners.

7 See Avicenna, *Canon* 1.1, and Haly Abbas, *Pantegni*, 1.24.

https://doi.org/10.1515/9783110612639-003

peutic task of physicians trained in complexionate medicine was to identify and monitor the complexions of their patients and to make adjustments to the regimen – the diet, medicines, bathing, exposure to celestial bodies, and other habits which affect the balance of our qualities.

Pietro d'Abano made significant contributions to complexion theory, and his analysis advances far beyond purely medical concerns. More than any other concept in his thought, complexion distinguishes the medical perspective on human nature from other anthropological discourse in theology or philosophy.[8] He seeks to define complexion as something far more significant than its qualitive (and thus contingent) status would suggest. It is a necessary quality, acting almost like a substantial form, in that a thing ceases to be what it is without the necessary complexion. Through complexion, he tells us, "we come to the understanding [of a thing], through the sense of touch, as though through its essential features".[9]

This material perspective (and I emphasize *perspective*, rather than materialism) did not need to conflict with other perspectives on the soul and its faculties – complexion could be a complimentary principle – but his ambitious use of complexion was certainly fraught, as it seemed so similar to Galen. It was commonly reported that Galen had defended the materialist position that equated complexion with the soul, certainly considered an unacceptable and dangerous claim. Pietro argues that Galen's position was misunderstood, while he stresses the Aristotelian origins of his interpretation of the theory, uniting the medical (Galenic) use of complexion to Aristotle's greater philosophical authority. One of Pietro's central concerns in his account of the concept is to defend the Galenic doctrine of complexion by arguing that Galen did not really equate complexion with soul, while simultaneously test the extent to which complexion can provide direct access to a person's soul and its operations.

Because the regulation of complexion was so central to both physical and psychological health, Pietro was also interested in whether there is a "best" complexion from the perspective of the soul. Like Roger Bacon, who had suggested that "[a] physician knows how to correct and improve complexions so that [his patients] are inclined to good and useful things for themselves and others, both in wisdom and morals",[10] Pietro suggested that the best complexion if properly understood could help perfect the soul and its virtues.

---

**8** Since I first argued for the centrality of complexion in Pietro's medical anthropology in my 2007 dissertation, now see also Chandelier and Robert 2013, and Jacquart 2014. On the concept of complexion in Antonio da Parma's medical anthropology, see Robert 2023.

**9** *Expositio*, 14.1, "In hac determinat problemata naturalia circa complexionem, in cuius quidem cognitionem venimus per sensum tactus, tanquam per essentialia. Ut apparet in *Tegni*: sunt enim signa sumpta ab ipso substantialia que necessarie sunt veritatis secundum Haly Rodoan."

**10** Bacon, *Secretum secretorum cum glossis et notulis*: "Et [medicus] scit per has vias rectificare et meliorare complexiones eorum ut inclinentur ad bona et utilia sibi et aliis, tam in sapiencia quam in moralibus . . ." (ed. Steele 1920: 9).

## Aristotle and Galen on Complexion

Although complexion theory had been a part of learned medicine since the translating efforts of the twelfth century, its importance increased dramatically in the decades around 1300. The most popular medical textbooks, including Johannitius' *Isagoge* and the *Pantegni* of Haly Abbas, treat complexion between the elements and the humors in the introductory sections devoted to the "naturals". However, twelfth-century physicians tended to regard the humors as the more important concept.[11] The greater emphasis on complexion in the later thirteenth century was part of the project to re-evaluate the connection between medicine and philosophy. Scholars turned their attention to more sophisticated descriptions of complexion in medical sources which had not been available earlier. These included Avicenna's *Canon*, Averroes' *Colliget*, and several Galenic treatises, notably Galen's *De complexionibus*, which had attracted little attention until now.[12] The main sources for Aristotle's doctrine on the subject were the *De partibus animalium* and the *Meteorologica*.

Study of these sources revealed a tradition of differences of opinion among the medical authorities, and more generally between uses of the concept by philosophers and physicians. Because physicians in this era saw complexion as one of the main concepts binding medicine to natural philosophy, they were understandably eager to explain differences between the Galenic and Aristotelian versions of the theory. Pietro d'Abano provided one of the most thorough attempts to address these differences. Without a doubt, one of his motivations for reconciliation was to unite Galenic medicine to Aristotle's far greater authority in ethical and psychological matters. Pietro devotes twelve long *differentiae* (17–28) to complexion theory in the *Conciliator*, spanning more than 67 columns in the printed editions, amounting to a significant treatise on the subject. This follows a group of closely related questions (*diff.* 11–16) about the elements, in the normal order of medical textbooks.[13]

One issue of concern for Pietro was that there appeared to be fundamental differences between Aristotle and Galen concerning the application of complexion theory to different types of bodies. Aristotle's strictest remarks on complexion in his *Meteorologica* seem to limit its explanatory power to elements and simple homogenous bodies, which in animals include blood, bone, flesh, and other fluids and tissues. According to Aristotle, all natural substances derive their characteristics from the four primary qualities – which may be better understood as dynamic forces – of hot, cold, wet and dry.[14] Hot and cold are active qualities, while wet (sometime rendered "fluid") and dry ("solid") are passive. The interaction of these qualitative forces causes unspecified mat-

11 For the blend of humors and complexion, see Wallis 2015.

12 Galen, *De complexionibus* (*Peri Kraseon*) (ed. Durling 1976).

13 For these questions on the elements, see Siraisi 1985.

14 Aristotle, *Meteorologica* IV.1 = 378b10–379b9 (ed. and transl. Lee 1952: 291–295); *De partibus animalium* II.1 = 646a13–13–20 (ed. and transl. Peck 1937: 107).

ter to form the four elements (or primary substances) of earth, air, fire, and water, each of which is the result of the combination of two primary qualities. Thus, the interaction of cold and wet qualities forms water, while hot and dry are the qualities of fire. The elements and their qualities combine to form what Aristotle calls homogenous or uniform (*ὁμοιομερής*) bodies, such as stone or dirt. The sensible properties (or secondary qualities) of uniform bodies, such as hardness, smoothness, or color, are determined by the interaction of the active and passive primary qualities.[15] For simple substances like these, complexion is practically identical with a substance's formal characteristics and can almost be treated as the "form" of an element or uniform substance.[16] This dynamic interaction is also the material or formal cause of all generation, decay, and other change. Aristotle calls the mixture of these qualities a "κρᾶσις".

Aristotle applied these physical principles to biology, teaching that the homogenous bodies are the uniform parts of the body, such as blood, bone, flesh, and other fluids and tissues.[17] The homogenous parts combine to form more complex heterogeneous parts, such as a hand, eye, or other organs; the heterogeneous parts, in turn, compose the entire animal organism. For these more complex bodies, Aristotle makes limited use of κρᾶσις (*krasis*). As part of the material explanation, κρᾶσις (*krasis*) is needed to understand elements and homogenous bodies, but insufficient to account for more complex organic bodies.[18] To describe the cold and dry qualities of a rock is to describe fairly completely the formal characteristics of the rock. However, to say that an eye, or other heterogeneous organ, is cold and wet, provides some information, but does not explain as much. A similar qualitative description of a dog or a human enlightens us even less.

Galen took these basic physical principles from Aristotle and applied them to medicine. In his *De complexionibus*, his most thorough treatment of complexions, Galen applies κρᾶσις (*krasis*) to homogenous fluids and tissues, but also to heterogeneous organs and entire organisms, narrowing the differences between these categories. He assigned appropriate complexions to each organ of the body (e.g. the brain is cold and wet), to different animals (e.g. bees are dry), ages (e.g. old age is cold and dry), genders (e.g. women are cold and wet), and many other things (see illustration 2, p. 51). Galen emphasized that these complexions are relative. A woman is not really cold and wet in an absolute sense, but usually colder and wetter than a typical man. Thus, "it is not absurd if the same body is said to be hot and cold, as long as it is not [in comparison to] the same object."[19]

15 *Meteorologica*. IV.8 = 384b24–385b5 (ed. and transl. Lee 1952: 337–341).

16 *Meteorologica* IV.12 = 389b23–390a10 (ed. and transl. Lee 1952: 370–371).

17 *De partibus animalium* II.1 = 647b10–21 (ed. and transl. Peck 1937: 117).

18 *Meteorologica* IV.12 = 390b10–390b14 (ed. and transl. Lee 1952 : 372–375).

19 Galen, *De complexionibus* (Peri Kraseon), 1.5: "Non inconveniens, si idem corpus calidum esse dicitur et frigidum, sed si et ad idem." (ed. Durling 1976: 23).

Galen also used complexions to account for the "formal" characteristics of living creatures, such as tendencies and capacities that most scholastic philosophers and theologians would attribute primarily to the soul. In doing so, he also gave the explanatory power of κρᾶσις (*krasis*) far greater latitude than had Aristotle. A suitable complexion for any plant or animal, he emphasizes repeatedly, is not some absolute amount of each quality, but the complexion that results in the best performance of its activities.[20] The best complexion for a fig tree is the one that produces the best figs. A dog's best complexion should cause fierceness while hunting and docility when around its owners. A well-complected man will have a "mean of boldness and fear, of pity and envy, and of solicitude and negligence. Such a person will be lively, friendly, and wise."[21]

These differences between Galen and Aristotle, among others, had already been identified by Arabic physicians. Averroes pointed out that this difference becomes clear in the number of complexions they each thought to be actually possible in reality. Aristotle names only four possible combinations of qualities, each composed of one active (hot or cold) and one passive (dry or wet) quality. Galen, on the other hand, allowed for nine different complexions: in addition to the four "composite" complexions of Aristotle, each "simple" quality could dominate on its own, or the four qualities could be balanced. For Averroes, who is often critical of Galen as a philosopher, Galen's use of complexion is sloppy and relative, while Aristotle has a more philosophically rigorous definition, in which the qualities in a mixture always exist in an absolute sense.[22] For Avicenna, this difference is an example of the different goals of theoretical philosophy and a practical field like medicine.[23] Pietro, however, choses a different approach in reconciling this difference: he argues that Aristotle and Galen essentially agree about complexion and he blurs the distinction between medicine and philosophy.

As might be expected, Galen's view on the matter dominated the medical tradition, with almost everyone holding the theory that there are nine possible complexions. In the discussion of complexion in the *Colliget*, Averroes had pointed out that Galen and the physicians were making inappropriate use of complexion first by applying it to complex bodies and second by adding additional complexions. He considered the appropriation of the concept for medicine to be a minor error; he found it acceptable as long as everyone understood it was an imprecise use. He was more critical of Galen for the second error of creating new "simple" complexions – pointing out the physical impossibility of any two of the dynamic qualities remaining perfectly bal-

---

**20** Galen, *De complexionibus* (Peri Kraseon), 1.6 (ed. Durling 1976: 30–31).

**21** Galen, *De complexionibus* (Peri Kraseon), 2.1: "Corpore quidem utique talis eucratotatos homo: similiter autem et anima medius certe est audacie et timoris, et misericordie et invidie, et cure et negligentie. Erit utique qui talis bene irascibilis, amicabilis, et sapiens." (ed. Durling 1976: 52).

**22** Averroes, *Colliget*, 2.1.

**23** On the different perspectives of medicine and philosophy, see Siraisi 1981: 188–195.

anced and thus ruling out the existence of simple complexions. Pietro agrees with Galen that there are nine complexions, although, in his opinion, the complexion in which all four qualities are perfectly balanced is not possible in reality, and perhaps is not even imaginable.

## Pietro's Reconciliation

Pietro devoted *differentia* 19 to solving this issue, "Whether a simple complexion, hot, cold, wet, or dry, from half [of a composite complexion], is possible or not."[24] On the one hand, Pietro's own solution is straightforward: he agrees with Galen that complexion is a relative concept. However, he surprisingly argues that even Aristotle's definition is relative and that whether a body is "simple" or "complex" always depends on the perspective. Really, only pure potency (prime matter) and pure actuality (form) can be unequivocally called "simple" substances.[25] All other substances are a composite of these. At the same time, all substances are simple in comparison to the composite of body and soul in the human organism, the most complicated natural body, in which "all grades of being are found, from the simplest elemental matter up to intellectual being".[26] Everything may be called either simple or composite in comparison to one of these standards.[27] Thus even "the elements are composite with respect to prime matter and form, yet they are simple in relation to mixed bodies".[28] In other words (even if we may be unconvinced by this solution) Galen does not really contradict Aristotle.

Pietro's attempt to show that Aristotle and Galen are both correct is illustrative of his tendency to erase differences between medicine and philosophy by arguing the difference of perspective about what is really the same thing. At first sight, his approach might seem to be merely an application of Averroes' position that medicine

---

**24** *Conciliator*, diff. 19 (ff. 29v–31r): "Utrum sit ponere complexionem simplicem ex semisse, vel calidem, vel frigidum, siccam vel humidum, necne."

**25** *Conciliator*, diff. 19 (f. 29v, col. 2, H): "Entium autem quoddam est simplex omnino, aliud compositum omnifariam, nonnullum vero intermedium. Primum [i.e. simplex] quidem duplex existit: quoddam enim est potentia tale pura, ceu hyle, cuius existit esse in posse; substantiatur enim per posse . . . .Aliud vero actu, tamen velut causans primus; est enim actus primus purus, et immunis ab omni potentia."

**26** *Conciliator*, diff. 19 (f. 30r, col. 1, A): "Compositum quoque omnifarie corpus est maxime humanum. Ipsum nanque caeteris compositius existit corporibus, cum in eo gradus omnium reperiantur entium. A materia enim incipiens sicut caetera: haec prima per simplicia corpora transiens et commista usque ad ens porrigitur intellectuale."

**27** *Conciliator*, diff. 19 (f. 30r, col. 1, B): "Quae diversa comparatione simplicia dici ponent et compositia."

**28** *Conciliator*, diff. 19 (f. 30r, col. 1, A): ". . . unde elementa composita sunt respectu materiae primae et formae; sunt tamen simplicia corpora relatione mistorum."

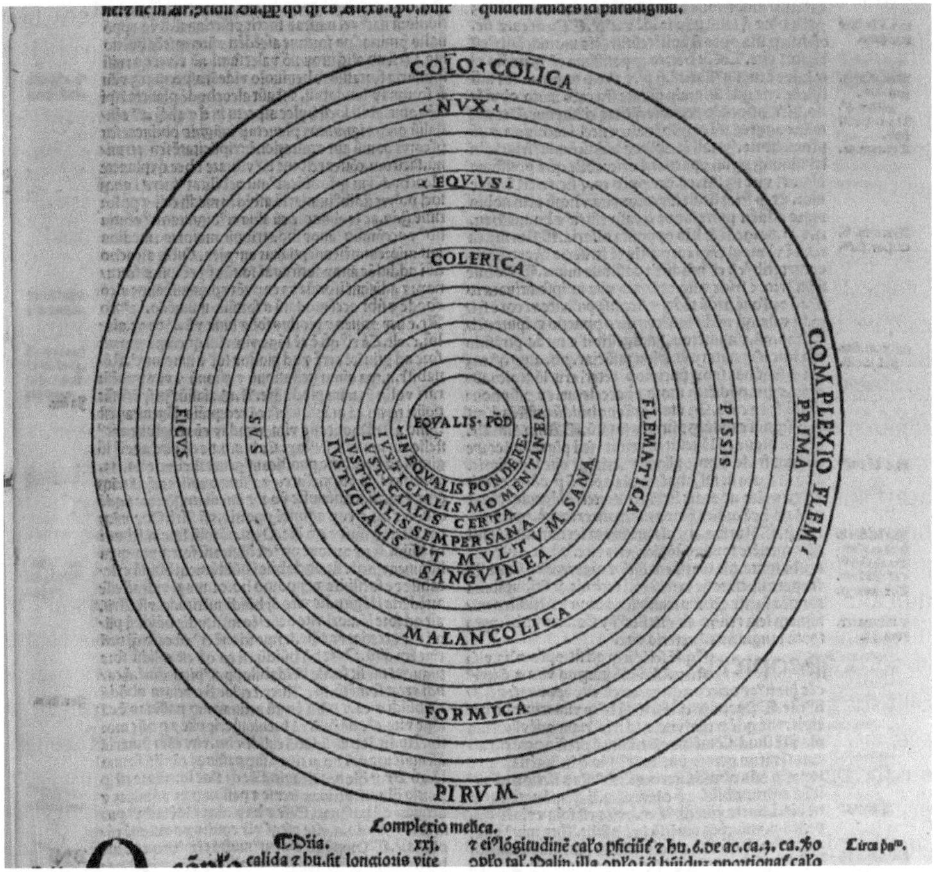


**Illustration 2:** This ring of complexion from the printed editions of the *Conciliator* includes the more or less balanced complexions in the middle rings, surrounded by the common imbalances and a few examples of the complexions of some animals and foods. Courtesy of the Countway Library of Medicine Rare Book Room at Harvard Medical School.

has a less precise definition that excuses the medical use of complexion. However, where Averroes distinguishes between an equivocal, medical use and an unequivocal philosophical use, Pietro argues that there really is no essential difference and that both depend on the perspective. With this bit of rationalization, Pietro addresses two central issues in the reconciliation of his authorities: 1) That the Galenic "simple" complexions are only simple in a relative way. That is, if a body is called simply "hot", this does not mean that its wet and dry qualities are absolutely balanced. In fact, it is actually impossible in most cases for this perfect balance to exist for more than a mo-

ment.[29] Rather, the mean between qualities should be understood imprecisely, like the approximate "mean" of a moral virtue.[30] More significant, but not explicitly stated, Pietro's explanation claims that 2) it is acceptable to apply the concept of complexion to all types of body, however "simple" or "composite" they are. If even elements are composite relatively speaking, we can also apply complexion to other, more complex, composite bodies.

His refusal to acknowledge a fundamental distinction between an absolute and relative meaning of complexion should be understood in the context of his radically "physiological" understanding of Aristotle.[31] This justification of the broad application of complexion is vital to his overall conception of Aristotle's and his own understanding of nature. When we consider the reconciliation of these in Pietro 's work, it is important to realize that, even if Aristotle limits the use of *krasis* in his more scientifically precise comments, in other circumstances, he uses qualitative explanations as a convenient way to explain the tendencies of complex organisms. The relation of the qualities of the heart and blood to behavior is most common in this regard. These have a powerful influence on the strength, courage and intelligence of an animal. For example, a strong animal has thick and hot blood; thin and cold blood results in an intelligent animal; thin and hot blood results in both courage and strength.[32] Furthermore, the mixture of the primary qualities can be held to account for many life functions: "It is surely sufficiently established that these four principles [i.e. hot, cold, wet, and dry] . . . are practically the causes controlling life and death, not to mention sleep and waking, prime and age, disease and health."[33] More generally, heat is indeed associated with soul throughout the biological works to the extent that evidence hardly needs to be drawn up – the primary seat of the soul and source of heat must be the same place.[34] Animals are classified, not just by hierarchy of the soul's capacities, but by amount of heat.[35] At the same time, in the pseudo-Aristotelian *Problemata* and *Secretum*, regarded as authentic by Pietro, these kinds of qualitative descriptions are common. Thus, for Pietro, Galenic complexion theory matched well with the "physiological" Aristotle. In sum, by arguing

29 *Conciliator*, diff 19 (f. 30v, col. 3, G): "Vero sciendum, quod huiusmodi complexiones quatuor simplices et maxime passivas non est ponere medias secundum mediatem arithmeticam consistentem in esse indivisibilis medii et veri. Quod ostenditur: quia si ponatur calidum frigidum excedere, medium contemperatum ex humido et sicco non poterit permanere . . ."

30 *Conciliator*, diff. 19 (f. 30r, col. 2, C): "Est medium aliud, quo ad nos habens latitudinem, et distantiam inequalem: ut 8 inter 6 et 12. Quale est medium assignatum in virtute morali (apparens *Ethicorum*, 2)." This is but one of many occasions that Pietro compares the strategies of medicine to ethics.

31 See Introduction: 5–6.

32 E.g., Aristotle, *De partibus animalium* II.2 = 648a5–14 (ed. and transl. Peck 1937: 119–121) and II.4 = 650b14–651a19 (ed. and transl. Peck 1937: 137–141). For places in Aristotle that connect physiology to intelligence, see Carbone 2010.

33 Aristotle, *De partibus animalium* II.2 = 648b5–8 (ed. and transl. Peck 1937: 123).

34 Aristotle, *De partibus animalium* III.5 = 667b22–28 (ed. and transl. Peck 1937: 249).

35 Pellegrin 1982/1986.

that Aristotle does not disagree with Galen, the Galenic application of complexion is philosophically legitimized.

## Complexion and Substantial Form

The first question about complexion in the *Conciliator* asks, "Whether complexion is a substance [or an accident]?"[36] In this question, Pietro's goal is first to refute the claim that Galen had ever taught that complexion can be identified with the soul, next to determine Galen's actual doctrine about the relation between complexion and soul and other substances or their forms, and then to explain his own position. Though directly known only by few outside of the medical faculties, Galen's work was widely reported as having identified the soul with the body's complexion.[37] Additionally, according to Averroes, Galen made the mistake of equating the immaterial intellective power with the cogitative faculty, an organic power based in the brain. These reports were sufficient for most scholars to dismiss Galenic psychology as incompatible with Christian doctrine.

Pietro is perfectly aware of Galen's reputation, acknowledging that "some modern authors claim that Galen maintained the soul to be a complexion".[38] He cites several examples of suspect passages from Galen, such as "nature, and soul, is nothing other than heat", from the *De tremore, iectigatione, et spasmo*.[39] Elsewhere he says that Galen speaks with less conviction saying that "either the complexion of the effective qualities is the soul, or the soul is altered by their complexion".[40] More frequently (and less problematically) Galen claims that "through following the complexions of

---

**36** *Conciliator*, diff. 17 (ff. 26r–27v): "Utrum complexio sit substantia, necne."

**37** For example, see Avicenna, *Liber de anima* I.2 (ed. Van Riet 1968: 43): "Quidam dixerunt quod anima est complexio." Pietro quotes Avicenna's text in *differentia* 17 of his *Conciliator* and takes for granted that it refers to Galen. Thomas Aquinas reports that Galen thought the soul is a complexion, which would mean it could not subsist without the body. See his *Quaestiones disputatae de anima* (ed. Bazán 1996: 8) q. 1, resp.: "[Q]uidam utrumque anime humane abstulerunt, dicentes animam esse armoniam, ut Empedocles, aut complexionem, ut Galienus, aut aliquid huiusmodi. Sic enim anima neque per se subsistere poterit, neque erit aliquid completum in aliqua specie vel in aliquo genere substantie, set erit forma tantum, similis aliis materialibus formis . . . [U]t probatur in II *De anima*; complexio autem et armonia qualitates elementares non trascendunt."

**38** *Conciliator*, diff. 17 (f. 27r, col. 1, A): "Quidam etiam modernorum imposuerunt Galeno velle ipsum animam fore crasim."

**39** *Conciliator*, diff. 17 (f. 26v, col. 2, H): "Ut *De tremore, iectigatione et spasmo* libro: "natura et anima non est aliud praeter calorem." Pietro cites additional passages from Galen in the list of initial objections (f. 26r col. 1, D–col. 2, A).

**40** *Conciliator*, diff. 17 (27r, col 1, A): "Aut crasis est anima effectivarum qualitatum, aut crasi earum alteratur."

the body, I demonstrated the virtues of the soul".[41] Taken together, Pietro admits that "these passages from Galen appear to sound as if complexion is a substance, and even as if it is the substantial form of the human body, namely the human soul".[42]

However, he argues that this appearance is deceiving; Galen never intended to formulate a coherent doctrine of the soul's substance and its relationship to the body.[43] Rather, Galen's investigation of the soul does not advance beyond observations of the soul's operations and the effect of complexion on these, which were relatively non-controversial subjects. In fact, Galen "plainly stipulates that complexion is something other than soul, saying that the humors and, more generally, the complexion alter the acts of the soul".[44] He relates Galen's story of a confrontation with a philosopher who was proposing "many opinions about the soul". Clearly dismissive of these opinions and perhaps the whole enterprise, Galen concludes that, on matters concerning the soul, "I leave this to those who judge of themselves, that they know this truly."[45]

If this conclusion settles the matter of whether complexion can be simply identified with the human soul, the issue was complicated by the fact that Galen *did* often seem to identify complexions with the substantial forms of other natural bodies, including individual organs, an apparent inconsistency. "It appears", Pietro says, "that the meaning of Galen is that complexion is the creator of the substances of the parts of an animal and that, therefore, it is the substance and nature of a body".[46] Further, he tells us that many of his colleagues in the medical *studia* accepted this interpreta-

41 *Conciliator*, diff. 17 (f. 27r, col. 1, B): "Concessum est ab optimis medicis et philosophis et a me per unum monimentum monstratum est, quod per crases corporis sequentes demonstravi animae virtutes."

42 *Conciliator*, diff. 17 (f. 27v, col. 2, H): "Propter secundum vero sciendum quod sermones Galeni praetacti apparenter sonare videntur quod complexio sit substantia, et tanquam corporis forma substantialis etiam hominis, ut anima humana."

43 There continues to be disagreement about the extent to which Galen intended to create a doctrine of the ontology of the soul. Pietro's interpretation of Galen's psychology is generally in agreement with that of Garcia-Ballester: "Galen . . . was not interested in formulating a doctrine on the soul-body relationship. Thus in the writings conserved to us – from the earliest to the latest – he went no further than affirm that there was some connection between the soul and body, but left the answer as what kind of connection existed between them rather vague." See Garcia-Ballester 1988: 119.

44 *Conciliator*, diff. 17 (f. 27r, col. 1, B): "Tertio tamen praefato videtur expresse velle complexionem aliud ab anima fore, dicens quod humores et universaliter crasis corporis alterant actus animae."

45 *Conciliator*, diff. 17 (f. 27r, col. 1, B): "[Philosophus] plures recitat de anima opiniones, post quas ait inquirere de anima, an eius substantia sit corporea. Neque cessat, quando corpus accidit alicuius rei animalis descendere in eam. Aut est aliquid corpori continuum, quod cum corpore decidit; aut spermatis est virtus, et positio aliqua. Et [recitat] cur conveniant cum creatione corporis accidentia animae. Relinquo hoc illis, qui de se aestimant, quod id sciunt veraciter."

46 *Conciliator*, diff. 17 (f. 26r, col. 2, B): "Propter primum vero sciendum, ut apparuit secundum sensum Galeni, quod complexio est factor substantiae cuiuscumque membrorum animalis. Complexio igitur est substantia corporis et natura eius."

tion of Galen and followed a similar usage.[47] These scholars write that, if the organ is understood as an "instrument of operation . . . they say complexion is the substantial form [of the organ]. For through [complexion] the organ operates and is brought into existence (*esse*) as an instrument; through this it is ordered towards act."[48] For physicians less intent than Pietro to use complexion to explain formal capacities, this interpretation was probably less fraught with potential controversy. Pietro, however, is eager to qualify this reading of Galen and distance himself from those who held it, calling it inconsistent and philosophically unsophisticated.[49]

Pietro argues that the identification of complexion with the forms of the organs is also a misinterpretation of Galen stemming from a lack of firsthand knowledge of his work and poor translations of what texts were available. Where Galen seems to say that the complexions of the body's organs are substances or substantial forms, it is because he often uses "substance" in a non-technical way. He further maintains that the works in which Galen most fully clarifies the relation of complexion to substantial form have not yet been translated into Latin. These include Galen's *De utilitate partium*, which was only available in abbreviated form, and *De optima compositione*. Pietro claims to have read the Greek texts and to have made his own translations.[50] He offers a sample from the *De optima compositione* that he thinks will clarify Galen's intent:

> Those things that exist first, are especially complexions, for they complete the proper substance of particular things, because bodies have this quality (hot, cold, dry and wet) and [bodies have their] nature on account of this [complexion]. It is in such a way that the "being" (*esse*) of flesh, nerves, and any other [organ] is made from the four aforesaid [qualities] of complexions. In as much as they are "natural bodies", I mean they are "complexioned things" and not "substances" or "bodies" *simpliciter*. And Galen adds: Thus, [complexions] exist in [natural bodies] *secundum rationem* of a substance.[51]

---

**47** Pietro perhaps has the teaching of the influential Bolognese physician Taddeo Alderotti in mind with this criticism. This is exactly the same distinction and argument made by Taddeo in his *Isagoge* commentary in response to the question: "Whether complexion is the substantial form of the members." In this question, Taddeo refers to the passages cited from Galen above as evidence that Galen thought "complexion is the substance of the organs" in contradiction to other authorities, notably Avicenna. See Taddeo Alderotti, *Expositiones in arduum aphorismorum Ippocrates volumen*, Venice 1527: f. 346.

**48** *Conciliator*, diff. 17 (f. 27r, col. 1, D): "Et sic dicunt complexionem formam substantialem fore. Per eam namque operatur membrum et in esse traducitur instrumenti; per ipsam namque ita ordinatur ad actum."

**49** Pietro may have had an interest in carefully distinguishing his own much more complex (and opaque) theories from those of Taddeo. Taddeo had unambiguously stated that the intellect belonged to the "animal" powers and was thus material. See Siraisi 1981: 210–213.

**50** *Conciliator*, diff. 17 (f. 27r, col. 1, C).

**51** *Conciliator*, diff. 17 (f. 27r, col. 1, C–D): "Est autem ipsa talis: "Existunt quidem primae ac maximae complexiones. Hae namque substantiam particularum complent propriam, quoniam hinc qualitatem habent caliditatem, frigiditatem, siccitatem et humiditatem corpora propter hoc et naturam. Est enim

Pietro hopes this passage will demonstrate that Galen clearly distinguishes the concept of complexion from substantial form. Although Galen typically avoids crediting Aristotle (despite the influence of Aristotle on his thought), Pietro takes Galen's substance (*ousia*) in this passage to be equivalent to the Aristotelian concept of (substantial) form, which is a reasonable interpretation. For both, the organ's substance or form is essentially its function.[52] In this passage from Galen, the complexion that exists prior to the formation of an organ determines its substance, as it simultaneously organizes the matter, and then accounts for its continued ability to fulfill its function. More to the point, however closely connected substance and complexion are, they are not identical concepts. In a "natural body", any union of matter and form, complexion cannot be identified solely with either aspect. In such a body, complexion exists "*secundum substantiae rationem*": it corresponds directly with the *ratio*, the essential features of the substance.

When we turn to Pietro's own account of the complexions of organs, we find that he draws heavily on this interpretation of Galen, simultaneously maintaining that complexion is different from substantial form while asserting their necessary relationship. Complexion, Pietro says, can be considered either as a thing or as a "mode".[53] It is clearly not a substance in the sense of an independently existing thing; but as a mode – the means by which the functionality and essence of the organ is preserved – complexion can be understood improperly as substance:

> Improperly, however, complexion is a substance. In as much as through substance, or substantial form, something is maintained in its complete essence; in the same manner, through complexion an organ is preserved in its essence and its operation remains.[54]

---

tale esse carni, et nervo, et aliorum unicuique hic, quod propter talem ex quatuor dictis complexionum factum est, prout sunt corpora naturalia, complexionata dico, et non substantiae simpliciter, aut corpora. Et [Galenus] subdit: 'Haec igitur ipsis secundum substantiae rationem existunt'."

52 On Galen's doctrines about the soul's operations in the organs, see Hankinson 1991.

53 *Conciliator*, diff. 17 (f. 27r, col. 2, A): "Propter tertium vero sciendum quod complexio potest dupliciter considerari: uno quidem quantum ad rem; aliter vero quantum ad modum. Siquidem primum dicendum impossibile omnino complexionem fore substantiam propter inconvenientia praetacta (*differentia* 13). Simpliciter eam dicendum accidens, quod est qualitas, existere, quia complexio ex elementis educitur, sive ex eorum substantiis aut qualitatibus. Cum enim in elemento sint duo, substantia et qualitas, proportionabiliter et in elementato debent duo existere. Non quidem ex substantiis elementorum consurgit complexio, quia id quod educitur ex eis, per compositionem elicitur, et talis secundum Avicennam est forma misti. Igitur ex qualitatibus educitur, quod alternatione perficitur."

54 *Conciliator*, diff. 17 (f. 27r, col. 2, B): "Si vero complexio non rem substantiae importet, verum modum: sic utcunque potest tolerari, improprie tamen, complexionem substantiam fore. Secundum enim quod per substantiam, sive formam substantialem, aliquid sustentatur in esse perfecto, ita suo modo membrum per complexionem in esse ipsius sustentatur et permanet operatio." The text continues: "Unde, quia illud a quo immediatius videtur actio provenire, manifestius est nobis, sicut instrumentum, ut complexio, quam illud a quo mediate, puta agente principali, ut substantia, cum per operationem in cognitionem alicuius perveniamus."

It is likely that Pietro's explanation of the exact nature of the relationship between complexion and substantial form did not completely satisfy his readers, then or now. He appears to lack the vocabulary to formulate a more precise description – indeed, it would probably have been unwise for him to offer a more unequivocal statement, given his trouble with the authorities. Instead, he offers the rather imprecise formulae we have seen or other vague suggestions of the immediacy of the bond between complexion and soul;[55] when it comes to the soul per se, he expresses approval of Galen's protestations of ignorance. Whatever his explanation lacks, it is apparent that he hopes to underline the point that complexion provides the natural philosopher with the most direct access to form; as such, it is an indispensable tool for psychology.

## The Best Complexion

One of Pietro's most novel contributions to the physiology of the soul is his thought on the connection between the best complexion and intellectual virtue. Regarding this, there are two closely related issues: what type of complexion was the "best" and how this complexion was connected to human perfection. As we have seen, Galen theorized that the ideal complexion for any living creature is the complexion that best enables it to fulfill its highest nature, which for a human is the rational faculties. He further proposed that, among all creatures, the human complexion has the most equal balance of the four qualities; therefore, the best human complexion is also the most balanced.[56] Avicenna agreed with Galen, adding that our complexionate balance is almost supernatural: God had given man had a more balanced complexion than could otherwise be found in this world.[57] This Galenic theory again posed challenges for Pietro: Galen's balanced complexion contradicts Aristotle's view that all living creatures are characterized by hot and wet qualities, and humans even more so. To add to the confusion, the Aristotelian *Problemata* (30.1) suggests that the cold and dry complexion of melancholy produces the best intellectual abilities and must therefore be best in some sense. Once again, Pietro manages to explain that these positions actually agree.

Pietro discusses different aspects of the best complexion in *differentiae* 18–20 of the *Conciliator*. As a physician, he could hardly be expected to disagree with the weight of

---

55 As he puts it at the end of this differentia, *Conciliator*, diff. 17 (27v, col 1, F): "Quando risibilitas aufertur, aufertur homo; sed non primo, sed ablatione rationalis. Ita et complexio aufertur, secundo forma substantiali sublata, ut anima." "When the ability to laugh is gone, the human is also gone, not primarily, but because the rational capacity has been removed. In the same way, when substantial form, that is the soul, is gone, secondarily complexion is removed."

56 Galen, *De Complexionibus* II.1 (ed. Durling 1976: 3–4).

57 Pietro often quotes this passage from Avicenna favorably: "[Deus] donavit homini complexionem temperatiorem quam in hoc mundo foret possibile invenire . . . ." e.g., *Conciliator*, diff. 5 (f. 9r, 2D).

Galenic tradition. And indeed, he is clear that the optimal state for a human is to be balanced. However, he claims that Aristotle is also correct because "the nature of the entire universe tends toward hot and is inclined toward humid".[58] The celestial bodies are ordered toward hot and wet, conferring these qualities especially to man among all living creatures. Thus, we are "balanced" because our complexion is most in conformity with the cosmos: man is a "paradigm of the universe". We are simultaneously "hot and wet" and "balanced", and this apparent difference is quickly dismissed.

But what physical form would a balanced complexion take? According to Pietro, there are two ways to understand it. One type of balance is the *complexio equalis quo ad pondus,* which results from the absolute equality of the "measure" or "weight" of the material elements in the mixture, so that no single element or quality surpasses the others.[59] The complexion *ad pondus* is simply not possible in nature.[60] Alternatively, balance may refer not to any absolute equality of weight, but to the "measure and qualities of the elements sufficient to carry out the [organism's] actions."[61] He follows Galen, and calls this the complexion of "justice" (*complexio equalis quo ad iustitia*).

To return to the melancholic complexion, *differentia* 20 of the *Conciliator* asks whether it may be possible that melancholy is actually closer to the absolutely balanced complexion of weight than the just complexion is.[62] He concludes that this cannot be the case: the cold and wet qualities of melancholy are simply incompatible with the hot and wet "balance" of the just complexion. According to his hierarchy of good or healthy complexions, even all the other imbalances (sanguine, choleric, and phlegmatic) are closer to a perfect balance than melancholy (see illustration 2).[63]

Beyond the *Conciliator*, in his commentary on the *Problemata,*[64] Pietro deals directly with the Aristotelian text that associates excellence in "philosophy, politics, po-

58 *Conciliator*, diff. 20 (f. 31v, col. 1, E): "Sic itaque natura totius universi in calidum tendit, in humidum inclinatur."

59 *Conciliator*, diff. 17 (f. 26r, col. 2, C): "[D]upliciter possumus imaginari evenire: aut enim quantitates et qualitates pondere et mensura elementorum sunt in misto aequales proportionatae non superantes, neque superatae. Qualitasque resultans ex eis sit media complexio interomnes extremitates simplicis qualitatis et compositae, ut caliditatis, frigitatis, et reliquarum. Caliditatis et sicci et caeterum et denominatur 'aequalis quo ad pondus.'"

60 *Conciliator*, diff. 18 (f. 28r, col. 2, C): "Propter tertium quidem primitus sciendum quod complexio ponderis non reperitur realiter in natura . . ."

61 *Conciliator*, diff. 17 (f. 26r, col. 2, D): "Complexio vero ad operationes relata temperamentum dicta 'iustitia' est, quae provenit quando tantum fuerit de mensuris et qualitatibus elementorum, prout suscipienti ad eius actiones peragendas."

62 *Conciliator*, diff. 20 (ff. 31r–33v): "Utrum melancholicus sit propinquior equalitati ponderis vel mensure temperamento iustitialis certi, necne."

63 *Conciliator*, diff. 20 (f. 31v, col. 1, G).

64 On this commentary, see Introduction: 11–13. For Pietro's defense of Aristotle's authorship, along with Pietro's argument for the philosophical import of this text, see Chapter 6: 111–114.

etry, and the arts" with melancholy.[65] Here, he severely qualifies the thesis in the text that excellence occurs in conjunction with a melancholic imbalance – instead he again maintains that a temperate nature is more appropriate for the use of reason, despite the suggestion of the base text. He observes that the physiological extremes produced by black bile more likely result in mania, incontinence, irascible vice, and bestial concupiscible excesses. Other melancholics rave because their reason is impaired.[66] True melancholics cannot reason and cannot conduct themselves well. He concludes that Aristotle could not possibly have meant to say what the text seems to say. Rather, what Aristotle intended by the type of melancholy that causes excellence in science and art is "not an excess" of qualities but is a "kind of balance" (*quasi ad medium temperamentum aliquod*).[67] From his medical perspective, Pietro seems to find the notion that the highest virtues could be best achieved through an extreme state of excess to be unappealing.

From the just complexion, on the other hand, "perspicuity of the senses will arise", from which will come "aptitude of the mind".[68] In short, from their physical qualities, people "are rendered intelligent or unintelligent". Because the human complexion is the most temperate, we are also the most prudent and most wise of all animals. A balanced complexion is needed for prudence because it "judges any excess more perfectly, since a mean is needed to hold right judgment and not to be inclined toward one side of an opposition" and sense excels with "a certain mean of the excellent qualities".[69]

---

**65** *Expositio*, 30.1: "Propter quid omnes quicumque excellentiores fuerint viri aut secundum phylosophiam, aut politicam, aut poesim, aut artes, videntur melancolici?"

**66** *Expositio*, 30.1: "[S]i huiusmodi melancolia fuerit superexcedens quantitate et caliditate, qui ipsa afficiuntur, redduntur furiosi, maniaci (quasi anima carentes, quia ratione carent, que est aliarum virtutem magistra), ac excessivi, prosilentes celeriter in actus et incontinentes, ac etiam irrogativi et curiosi . . . .Moventur etiam de facili in iram et concupiscentias brutales, ut tactus et gustus, sicut apparuit [in *Problema*] 28. Aliqui vero redduntur etiam verbosiores propter spirituum agitationes, rationis impeditivos . . . ."

**67** *Expositio*, 30.1: "Dicendum quod Aristoteles nititur declarare quod melancolici, cum naturaliter superhabundat colera nigra – non tamen superflue – ceteros preexcellunt in scientiis, et universaliter in omnibus artibus et actibus."

**68** *Conciliator*, diff. 18 (f. 29r, col. 1, D-col. 2, A): "Rursus cum homo sit ultimus finis et complementum universi a natura intenti (diff. 5), oportuit ex talibus esse, ex quibus posset caetera cognoscere et eis per ingenium et sapientiam dominari. Hoc autem erit, si in aliquo eius optimum ponatur miscibilium temperamentum. Ex his enim complexio consurgit temperatissima, ex qua et sensus perspicax elicietur. Ubi autem id, et aptitudo aderit mentis. Iuxta illud, secundi *De Anima*: molles carne, apti mente; duri autem inepti. Ex hoc enim redduntur ingeniosi vel non ingeniosi, et ideo dicendum hominum fore prudentissimum ac sapientissimum animalium." See also *Conciliator*, diff. 20 (f. 32r, col. 1, C–D).

**69** *Conciliator*, diff. 20 (f. 31r, col. 1, D): "Adhuc illa temperatior est complexio, quae perfectius de quolibet iudicat excessu, cum iudicem rectum oporteat medium omnifariam tenere et non ad aliquam contradictionis partem inclinare . . ., cum sensus in quadam extet mediocritate qualitatum excellentium."

Many of us can hope to achieve at least a moderately balanced complexion, at least within the range of imbalance that still counts as health.[70] However, Pietro's description of the very best type of complexion does not leave us very hopeful about attaining it. It is extremely rare and only happens when, through a perfect conjunction of the planets, someone is born who is halfway between angel and wise men.[71] On several occasions, he mentions Christ as someone who had such a complexion (at least in his human nature), but he hastens to add that the creation of Christ's nature was a divine act.[72] At any rate, few people can have this; the rest of us make do with lesser complexions and regulate our regimen in hope of attaining greater balance.

The fact that we can alter even our innate complexion with regimen is the crucial point for Pietro's medical anthropology. He distinguishes between our flowing (*fluens*) complexion, which obviously changes, and innate complexion, unique to every individual, which prepares for the introduction of the soul. This mutability might seem at odds with something so closely linked with the soul as substantial form. However, he insists that "although it may seem strange to many, innate complexion is able to be completely changed".[73] It is easiest to change one component of the complexion – e.g. if someone is hot and dry, he can be changed to hot and wet. It is much more difficult to change both qualities, so that a melancholic would have a very difficult time becoming sanguine, but it is possible. We can improve ourselves, physically, mentally, and morally by improving our complexion or the complexion of instruments of the body. More importantly from a professional standpoint, the physician can be the mediator of this change.

---

**70** *Conciliator*, diff. 20 (f. 32r, col. 1, A).

**71** *Conciliator*, diff. 18 (f. 28v, col. 2, G): "Est et alia complexio latitudine participans aliquantula, dicenda quidem angustior valde rarissime inventa. Neque enim duo ipsam habentes, nedum plures possunt in uno tempore existere, cum ad id videatur maxima coniunctio, puta Saturni et Jovis in principio Arietis cum aliis et aspectus omnis supercoelestium operari natura iubente divina, ceu quando quis insurgit eadem consignante prophetia legem imponens novam sapientes dogmatizans et homines. Est enim hic medius inter angelos et sapientes nedum homines, secundum Isaac, Amaraan, Avicenna, et Algazelem. Et talem fuisse dixerunt quidem Moysen et Machabei mendaces, Christum, Mariae filium."

**72** *Conciliator*, diff. 20 (f. 32r, col. 1, C): "Christum tamen testantur hierarchae temperatissimum extitisse, quod actus divini edocuerunt eiusdem, inquantum autem natura participavit humana."

**73** *Conciliator*, diff. 22 (f. 35r, col. 2, B): "Propter tertium autem sciendum, licit pluribus videatur extraneum, quod complexio innata permutetur omnino." Pietro was not alone in suggesting the mutability of innate complexion: see Siraisi 1981: 258–261.

# Chapter 3
# *Spiritus*, Life, and the Generation of the Soul

Because of its fundamentally medical structure, the *Conciliator* does not contain any questions that are directly about the soul *per se*.[1] However, scattered among various topics, we find many lengthy digressions about the nature of the soul, of the sort that Avicenna had argued are beyond the requirements of medicine.[2] These topics include the number of principal parts in the body, the vital virtue, respiration, reproduction, and the nature of heat and spirit.[3] The most important entity in these discussions is certainly spirit (*spiritus* or πνεῦμα), the corporeal, but extremely rarefied, substance that carries out the soul's activities. Spirit is the most important vehicle for the distribution of complexions throughout the body. Like complexion, spirit is also a very flexible concept, accounting for a host of actions whose source is otherwise invisible. Pietro emphasizes its power and flexibility, calling it alternatively the *delator* (informer) for the soul, a "wild horse tamed by reins", or a candle illuminating the body.[4] It is essential for the formation of the embryo, for sensation and motion, for vivifying the powers shared by body and soul, and for being the closest material substance to the soul. In these discussions, Pietro pushed the limits of material explanation, emphasizing the material aspect of the union of body and soul to such an extent that it is no surprise that it attracted the attention of critics. Pietro's discussion of the formation of the embryo is almost animistic in its focus on the causal role of innate spirits, carrying their mixture of celestial and elemental heat, in preparing for the arrival of the soul and enabling its activities. After first considering the nature of spirits, the rest of this chapter examines their role in the origin of life and generation of the soul.

## *Spiritus*

Arabic Galenism transmitted the idea that there are three types of spirits in the body: the natural, vital, and animal, although Galen's own writings only seem to include the

---

1 The second half of this chapter, on the origin of the soul, is a revised version of Klemm 2012b, in a volume published by Brill. I thank Brill for permission to reproduce it here.

2 Avicenna, *Canon* 1.1.5.5. On the soul as an object of investigation in natural philosophy, see Zupko 1997, and De Boer 2013.

3 For some other aspects of the soul not touched here, including nutrition, vision, and touch, see Hasse 2001.

4 *Conciliator*, diff. 38 (f. 59v, col. 2, G).: "[S]piritus efficitur aethereus, qui organum existit ad actum, velut equus agrestis admonitus a freno." Also, *Conciliator*, diff. 48 (f. 72r, col. 2, A): ". . . spiritus per vias illius velociter transire, cum oporteat ipsum celeriter totum corpus irradiare, ut illuminans domum velut candela."

https://doi.org/10.1515/9783110612639-004

vital and animal.[5] The *Isagoge* briefly describes the three types and their source: the natural spirits are found in the liver, the vital spirits in the heart, and the animal (or psychic) spirits in the brain.[6] The *Pantegni* includes the same three types of spirit and source, and offers a little more detail.[7] The natural spirits arise in the liver from the purification of all the humors, travel in the veins through the whole body, and govern the natural powers. The vital (also called "spiritual") arise in the heart, travel through the arteries, and govern the vital powers. The animal spirits are generated at the base of the brain in the so-called *rete mirabile* from vital spirits rising from the heart.[8] However, overall, there is very little space in the *Pantegni* devoted to spirit compared to the other categories of naturals – just one short chapter at the end of Book 4. As the text begins to describe the role of the animal spirits in sense, memory, and reason, it abruptly concludes with the remark that this is a subject best left to philosophers.[9]

Several philosophers in the twelfth century accepted the suggestion in the *Pantegni*, moving beyond the strictly physiological understanding of spirits to explore the relationship between the material spirits and the immaterial soul and analyze the place of spirit as a causal entity.[10] Inspired by Qustā ibn Lūqā (ca. 820–ca. 912), *De differentia animae et spiritus* (On the Difference Between Soul and Spirit, translated ca. 1140), spirit was often discussed as a special kind of substance, essentially different from ordinary matter, sometimes endowed with celestial, almost divine qualities. However, James Bono argues that this twelfth-century tendency faded by the middle of the thirteenth century, with Albertus Magnus and others bringing spirits firmly back to the material realm, perhaps seeking to rid spirits of animistic or pantheistic possibilities, while still arguing their importance as a medium of the soul. Bono further suggests that this picture of a late thirteenth-century consensus about spirit might be revised if medical philosophers – such as Pietro d'Abano, in particular – are taken into account.[11] Indeed, like the concept of complexion, spirits became increasingly important among philosophically inclined physicians around 1300. By this time, Avicenna had become the most influential source on spirits.[12] He had revised the descriptions in the *Pantegni*, which had attributed rational powers to the animal spirits

5 For medieval discussion of spirits, see Bono 1984. On Galen's position, see Temkin 1951.

6 *Isagoge*, sec. 17, (ed. von Maurach 1978: 155).

7 *Pantegni, Theorica*, 4.19 (ed. Burnett 1994: 113–120), which includes the two versions of the chapter on spirits which circulated.

8 On the Galeninc *rete mirabile*, which does not actually exist in humans, see for example Prioreschi 1996.

9 Bunett, ed. 1994: 11: "Sed quia disputatio huiusmodi non huic suppetit intentioni cum philosophica sit, hic dimittenda censuimus ista."

10 See Chapter 1: 27, and Bono 1984: 98–112.

11 Bono 1984: 127.

12 Jacquart 2013: 600–601. Avicenna's most significant discussion of the spirits is in the *De viribus cordis* (translated ca. 1178) which circulated with Avicenna's *De anima*. This is included as an appendix to Van Riet's edition of *De anima* 1968: 2.187–210.

in the brain, to be more compatible with theological doctrines by maintaining the immaterial nature of the rational soul. Because he assigned to the animal spirits most of the sensitive functions that Galenic medicine attributed to vital spirit, this left vital spirit without a clear place in the system of spirits. Avicenna's influence is apparent in the work of other late thirteenth-century physicians who sought to understand how the vital spirits fit into the standard tripartite division of the soul.[13]

Pietro's most complete descriptions of *spiritus* in the *Conciliator* are found in *differentia* 59, where he questions whether spirit and heat are the same thing, with a appended question which asks whether spirit remains one and the same during the life of an animal, and *differentia* 38, which asks whether there are multiple principal parts of the body.[14] His definition of "*spiritus*" is inclusive and sprawling (as his definitions often are) recognizing that it has a wide range of meanings beyond the strictly medical: spirit is part of the trinity, the entity in the expression "spirit blows where it wishes", the vehicle of "demonic favor and depravity", as all those things meant by the phrase "spirit is willing, but the flesh is weak", as wind, or breath, for the subtle bodies sublimated by alchemists in the transmutation of metals for production of the elixir called the philosopher's stone.[15] Or, finally arriving at the medical definition, spirit is the subtle, airy, body which is the bearer of the soul's powers.[16] As is often the case in his wide-ranging definitions, it is unclear to what extent Pietro regards these different meanings of "spirit" to be compatible with each other, or with medical spirits. Certainly, he intends to suggest that some of them are. As for medical spirits,

> The ancients agree that spirit is a subtle body proceeding from the heart in the arteries to the whole body, and it makes it have life and breath and ascends to the brain and proceeds from there into the nerves through the whole body and gives it sense and motion.[17]

These subtle bodies are almost immaterial:

> Spirit is a body that also has incorporeal being in some way and is a mediating bond between extremes [i.e. body and soul] having opposed substances. When it is in good condition (in so far

---

**13** Siraisi 1981: 166–169.

**14** *Conciliator*, diff. 59 (ff. 87v–88v): "Utrum calor et spiritus sint idem, necne?" diff. 59 appendix (ff. 88v–89r): "[An] spiritus permaneat idem numero in omni vita animalis?" and diff. 38 (ff. 58v–60r): "Utrum plura sint membra principalia, necne?

**15** *Conciliator*, diff. 59 (f. 87v, col. 2, F): "Spiritus siquidem multiplicitur dicitur, sumitur nanque pro una trium personarum divinarum, iuxta illud Johannis, 'Spiritus, ubi vult, spirat;' et pro substantia daemoniaca gratificata, vel depravata; et pro omnia secundum illud evangelii: 'Spiritus quidem promptus, caro autem infirma;' et pro vento (*Meteo*, 2); ac pro anhelitu (ut *Pronosti*, primo); et pro corporibus alchimicis sublimatis subtilibus metalla transmutantibus, ex quibus elixir, idest, lapis philosophorum construitur."

**16** *Conciliator*, diff. 59 (f. 87v, col. 2, F): ". . . ac pro corpore subtili aereo virtutum latore . . . ."

**17** *Conciliator*, diff. 59 (f. 87v, col. 2, G): "Convenerunt antiqui, quod spiritus est corpus subtile procedens a corde in arterias ad totum corpum, et facit ipsum habere vitam et anhelitum per pulmonem et ascendit ad cerebrum et procedit ab eo in nervos per totum corpus et facit ipsum sensum et motum."

> as it is of such or a certain type of quality), it provides the soul with its entire means of operation and is the substance of its affections.[18]

When spirits are "very subtle, pure, and splendid", they are "delightful" to the rational soul.[19]

Sometimes Pietro states that there are three types of spirits, with some spirit descending to the liver where it becomes natural spirit; however, he usually attributes this opinion to the physicians alone.[20] Instead, "more properly speaking, there are only two types of spirit", in the heart and in the brain, "but in the liver there do not reasonably appear to be spirits".[21] Most of the time, he is not particularly interested in this issue of whether there are two or three types of spirit. More often he simply refers to *spiritus* in general. He is much more interested in the difference between the innate spirits and the flowing spirits.[22]

Flowing spirits are spirits which are produced by the heart after it is fully functioning in the embryo (as is explained below). These, like the flowing complexion, are quite mutable: "The flowing (or running) spirits, as the name suggests, do not remain the same always."[23]

Innate spirits, however, are present from the moment of conception before the formation of any organs, and are a permanent feature: "Innate spirit, implanted primarily in the heart and other parts with its virtue, remains one and the same in number continually."[24] They must be permanent, he argues, because they are so close to

---

18 *Conciliator*, diff. 38 (f. 59v, col. 2, H): "[Spiritus est] corpus utcunque incorporeum ens, et ligatio media existit extremitatum substantiam contrarium habentium. Hic quippe bene consistens, dat animae omnia secundum rationem operari, secundum quod tali huiusmodi fuerit qualitate, ceu affectus substantia, ut in editione *Physionomiae* declaravi."

19 *Conciliator*, diff. 38 (f. 59v, col. 2, G–H): "Si vero subtilissimus, et puris imus, splendidissimus, anima diligitur rationali."

20 *Conciliator*, diff. 38 (f. 59v, col. 2 G–H): "Spiritus ex corde per arterias venales demandatur ad cerebrum, qui prius digeritur, et subtiliatur in replicatione quadam sub cerebro dicta rete mirabile et nunc spiritus animalis efficitur vitam potens praestare et sensum cum motu, ubicumque expanditur. Eodem quoque modo a corde spiritus demandatur in hepar, qui in eodem permutatur in naturalem. Secundum vero medicorum plures, spiritus naturalis in hepate creatur . . ."

21 *Conciliator*, diff. 59 (f. 87v, col. 2, G): "Nosce quoque, quod propius loquendo non sunt nisi duo spiritus, vitalis scilicet, et animalis. Unde *Colliget* Primo: Spirtus sunt duo, scilicet spiritus apparens in corde et spiritus apparens in cerebro, sed in hepate non apparet per viam sensus esse spiritus."

22 *Conciliator*, diff. 59 (f. 87v, col. 2, G): "Qui duplex est innatus complantatus, qui proprius est substantia corporea ex elementtorum commistione generata vergens in coelestium similitudem corporum (differentia 48), cum reliquis ibi recitatis. Et pro fluenti suscipitur magis, seu currenti, ceu ibidem ostensum qui est vapor sanguinis temporatus (7, *De Iuvamentis*), seu subtilitate secundum aliquam constructionem, cuius cor extat principium, de quo Isaac diffinitionum."

23 *Conciliator*, diff. 59 (f. 88v, col. 2, G): "Fluens vero, seu currens, ut nomen denotat, non permanet idem semper, verum alius efficitur . . ."

24 *Conciliator*, diff. 59 (f. 88v, col. 2, F): "Sciendum quod spiritus innatus complantatus principalius in corde caeterisque partibus eius virtute permanet numero unus, idemque contiue."

the soul, that they are practically the formal part of an animal. Of "the three substances present at the beginning of our generation, spirit, blood, and airy moisture", spirit is like form in the hylomorphic union of matter and form, while the others are matter.[25] Just as "the soul remains one and the same continually", the spirit must also, since it is "the first appendage (*appendiculum*) of the powers of the soul".[26] The spirit "quasi-informs (*quasi informat*)" the body like a "formal part (*pars formalis*)".[27] Innate spirit is able to remain permanent and not be diminished, because it is "nourished by the flowing spirit", in the same way an organism remains one and the same by being nourished throughout its life.[28]

Responding to his question of the relationship between spirit and heat in *differentia* 59, he explains that, although spirit is the unique vehicle for the body's heat, heat and spirit are not exactly the same thing and can be discussed separately.[29] He clarifies that the heat he is talking about is not the elemental heat of fire, but innate heat, a temperate and light heat. This innate heat is a composite of elemental and celestial heat, seated in the heart and diffused through the other parts of the body. Innate heat and spirit make up the same substance but are not identical.

Although it is almost immaterial, Pietro explains that spirit "is a corporeal substance, generated in the heart from the subtlety of the humors or from their vapors".[30] It is created by processes no more mysterious than the ordinary digestion of matter entering the body, breathing, and the pumping of the heart:

> The 'food' of which the spirit is constituted is air. Entering the heart, and [becoming] more subtle by breathing, [the spirit] ascends through the arteries to the brain.[31]

---

**25** *Conciliator*, diff. 59 (f. 88v, col. 2, G): "Quod autem innatus permaneat idem numero continue, fulcitur ratione: quoniam tria corpora circa principium generationis nostrae complent substantiam: spiritus, sanguis, et aerea humiditas."

**26** *Conciliator* diff. 59 (f. 88v, col. 2, G): "Adhuc anima permanet una eademque continue, sed his spiritus creatus est primum appendiculum virium animae."

**27** *Conciliator* diff. 59 (f. 88v, col. 2, G): "Rursus si permanet illud in quo aliquid inseparabiliter fundatur et id permanet ceu sunt materia et forma. Sed humidum radicale irrestauratum . . . spiritus autem innatus ipsum complectitur, et quasi informat. Itidem illud quod proportionabiliter se habet, ut pars formalis, et solida, permanet idem numero continue (ut visum diff. 56), sed talis est spiritus, ut tactum."

**28** *Conciliator* diff. 59 (ff. 88v, col. 2, G–89r, col. 1, A): "Amplius id permanet semper idem numero, quod nutritur; nutritio enim in permanentia ordinatur animalis, sed spiritus innatus nutritur ex fluenti (diff. 155)."

**29** *Conciliator* diff. 59 (f. 88r, col. 1, D): "Calor et spiritus nec sunt idem realiter differentes sola ratione, ne duo differentia corpora . . ."

**30** *Conciliator*, diff. 59 (f. 88r, col. 2, C–D): "[S]piritus . . . est substantia corporea ex subtilitate humorum, vel eorum vaporibus, secundum aliquam complexionem in corde generata calor ens."

**31** *Conciliator*, diff. 38 (f. 59v, col. 2, G–H): "Cuius quidem spiritus aer cibus constituitur. Inspiratione enim ingrediens cor subtiliatus per arterias ascendit ad cerebrum. Quam quidem nutritionem animo concedere, Galen cum Haly, etsi Aristoteles et Averroes displiceat et Avicenna (diff. 155). Qui siquidem spiritus corpus existat, configuratur corpori. Si vero subtilissimus, et puris imus, splendidissimus,

"Galen and Haly Abbas", he comments, "grant this nutrition to the soul, even if it displeases Aristotle, Averroes, and Avicenna." Because "if indeed spirit is a body, it is composed of body" and thus can be altered by diet and regime.

In *differentia* 38, which considers the number of principal parts of the body, Pietro discusses the relationship between spirits, the heart, and the nature of the soul, namely whether the latter is singular or plural. This is the appropriate place to discuss this, because, he says, anything we say about "the unity or diversity of the soul corresponds to unity or diversity of its parts" and spirits and the organs in which they are seated are the parts most associated with soul.[32] He offers three options for the way to understand soul taken from Aristotle's *De anima*:

1) the soul is a single innate essence from which all the virtues emanate;
2) the soul is a transcendent and extrinsic principle operating its virtues from outside the body;
3) there are many "souls" in one body represented by a collection of regent powers.[33]

He attributes the second possibility to Platonists, saying that they hold that the soul does not require the body and senses to know, except to be stimulated to remember what it already knows. But he dismisses this as an implausible idea.[34] The first and third possibility, he associates with Aristotle and medicine respectively.

These two accounts of soul, do not really differ, but are due to the difference in perspective of physicians and philosophers: philosophers look at the root principles of things, the *medici* look for the more manifest principles, as things appear now.[35] From one perspective (the medical), there are multiple principal parts; from the other

---

anima diligitur rationali. [Spiritus est] corpus utcunque incorporeum ens et ligatio media existit extremitatum substantiam contrariam habentium."

32 *Conciliator*, diff. 38 (f. 59r, col. 2, B): "Quidam sciendum, quod iuxta unitatem aut diversitatem animae, unitantur aut diversificantur partes, cum earum ratio sumenda sit a forma vel fine."

33 *Conciliator*, diff. 38 (f. 59r, col. 2, B–C): "Sciendum quoque de anima differentes fuisse opiniones, secundum Aristotelem, 2 *De Anima*:

1) nam quidam putaverunt animam fore unam essestiam facientem per se actiones suas sine instrumenti diversitate.
2) Putaverunt et alii quod anima sit sapiens per se ipsam sciens omni et non eget sensibus et instrumentis nisi ut per ipsa excitetur ad id, quod habet naturaliter in se ipsa. Quidam etiam dixerunt hoc fieri quasi reminiscendo, nam accidit ei interdum oblivisci, ut Platonici. Unde Boetius *De Consolatione*: quod si Platonis musa personat verum. Quod quisque discit immemor recordatur.
3) Nonnulli fuerunt dicentes animam non esse unam, sed multas, et quod anima, quae est in uno corpore animarum, extat collectio regentis, videlicit animae et apprehendentis et irsacibilis et concupiscibilis."

34 *Conciliator*, diff. 38 (f. 59r, col. 2, D): "Posito vero Platonis . . . annullatur sufficienter de anima, quare cum id instantis etiam non sit inquisitionis, suspendatur ad praesans."

35 *Conciliator*, diff. 38 (f. 59v, col. 2, F): "Ut verificetur aliqualiter dictum Galeni quod aliquid potest dici principium alicuius dupliciter: aut radicaliter et primo, vel manifestative et secundo."

(the philosophical), there is one.[36] The most true account is that there is only one soul, just as the heart is the first principal part.[37] However, the heart does not have all the powers of the soul in actuality, since it lacks the necessary instruments, but has only the vital powers with their spirits, which are in it radically and formally.[38] The animal powers are perfected in the brain and the natural in the liver. Overall, the reasoning is somewhat circular: because the soul is singular, there must be only one principal part; because there is only one true principal part (the heart, where all the spirits are produced), the soul must be singular. The heart is the single seat of the soul, where the soul is first inserted.[39]

## Life and the Generation of the Soul

According to Aristotle's *De anima*, soul is a substance in the sense that it is the form or actuality of "a natural body which potentially has life".[40] Aristotle further specifies that a body of this type must be a "body that has organs". The requirement that an ensouled body have an appropriate structure can be understood in a way that prioritizes matter – that is, what is most special about soul as a kind of form, is *not* some special ability of the soul itself, but rather it is the special kind of body to which it is united. After all, as Aristotle continues, the study of the soul must be done "in connection with the parts of the body".[41] Pietro addresses this topic of what kind of body meets the requirements to "potentially have life" in *differentia* 48 of the *Conciliator*, which asks whether the sperm settled in the womb is "animate". Does it have soul immediately after conception?[42] In other words, because a substantial form provides a thing with its identity, the question considers the difficult issue of exactly when the new entity in the womb becomes a human being. To answer these questions, Pietro offers a detailed account of the development of the embryo from the moment of con-

36 *Conciliator*, diff. 38 (f. 59v, col.2, F): "Primo siquidem modo est unum membrum principali; aliter vero plura."

37 *Conciliator*, diff. 38 (f. 59v, col. 2, F): "Prius quidem sicut cor, cum apud Aristotelis, una sit anima, saltem inquantum motus principium et sensus illi magis inditum, totum enim ipsius substantia in ipso fundatur."

38 *Conciliator*, diff. 38 (f. 59v, col. 2, F): "Non tamen omnes virtutes in eo prodeunt in actus, cum instrumentis careant determinatis, sed solum vitalis, quae extat radicaliter in illo et formaliter, ac etiam manifestative animalis vero virtus et naturalis radicaliter solum. Formatur vero animalis et perficitur in cerebro et naturalis in hepate, similiter apud Avicenna maxime *De Anima* 5."

39 *Conciliator*, diff. 38 (f. 59v, col. 1, G): "Itidem illud [i.e. cor] est principium membrorum, cui primo inseritur anima, et maxime motiva. Hoc autem est cor; primitus nanque omnibus vegetatur et movetur. Unde vitalis virtus aliorum omnium radix in ipsum configitur primo."

40 *De anima*, II.1 = 412a20–21 (ed. and transl. Hett 1936: 69).

41 *De anima*, II.1 = 412b17–18 (ed. and transl. Hett 1936: 71).

42 *Conciliator*, diff. 48 (ff. 71r–72v): "Utrum sperma decisum sit animatum, necne?"

ception until it has the structure and qualities to meet the material requirements.[43] Ultimately the soul will be infused directly by God; however, even divine action needs suitable matter.

The starting point for Pietro's response is the moment after conception when the mixture of sperm and menstrual blood in the womb acquires an identity independent of the parents, even if it requires nourishment from the mother. This new organism possesses its own spirits and is directed by its "informative virtue".[44] It is an independent thing with its own form, but is it a human being? The short answer is that it is not; it is only potentially so.[45] At this early stage, it is like the seed of a plant in the ground. With the right conditions for development and with no interference, it will eventually become a plant and shed its old identity as a seed. Thus, the embryo, like the seed, exists in a state of distant potentiality with respect to its final state, not in possession of soul, nor lacking the potential to acquire one.[46]

The more complete answer describes in great detail the transformation of the material from the initial mixture of the "vaporous foaminess" of the mother's blood with the heat in the sperm to the creation of the heart, liver, brain and umbilical cord, ultimately resulting in a body able to receive a soul as its new form.

This pre-animate development is directed by a principle called the "informative virtue", a power collected in the semen from both the soul of the father and from celestial bodies at the time of coitus, which supplies the embryo with its early form.[47] Pietro's usual insistence on the principle that a form needs to be united to suitable matter leads him to marvel at the activities of this "divine" virtue, which seems to defy this principle. The mixture of fluids to which it is united seems to lack appropriate instruments, yet it accomplishes so much:

> This divine power [the informative virtue] is nobler than all other powers of generation, because it is simple and almost immaterial. Not directly in contact with matter, it lacks a more distinct subject conveying it to give the appropriate shapes and forms of the organs. However, what it

---

43 For the first few days after conception, the entity is technically a "zygote", but I will refer to it as an "embryo" throughout.

44 *Conciliator*, diff. 48 (f. 71r, col. 2, B): "Decisum autem dicitur sperma, cum separatum fuerit a coeunte, in loco convenienti susceptum, spiritibus principalium fulcitum, ab informativa virtute possessum."

45 *Conciliator*, diff. 48 (f. 71r, col. 2, C): "Adhuc illud est potentia, quod in actum deducitur, nisi aliquid impediat. Tale quidem est sperma. Potentia nanque in actum naturaliter inclinatur, velut imperfectum ad perfectum. Quod si dicatur omnem potentiam in actum simpliciter duci, eo quod ociosa foret nisi perveniret in ipsum; nihil autem frustratur in rerum natura proprio fine."

46 *Conciliator*, diff. 48 (f. 71r, col. 2, D): "Nec enim animam habet actu, ut animal perfectum, nec est privatum ea, sicut mortuum, sed eam habet in virtute, longius tamen."

47 *Conciliator*, diff. 48 (f. 71v, col. 2, G): "Rector autem huius tam divini operis virtus est dicta informativa ab anima parentis decisa in actum per impulsionem coeuntis incitata."

> does have is spirit; for this subtle body is a conveyor of power. It also lacks an instrument, as it simple like the other powers of mixtures.[48]

The informative virtue does not have an instrument in the sense of a part of the body intended for its purpose. Rather, it is united to the spirit, which acts as its instrument. Like the capacities of all uniform bodies, the powers of the spirit can be sufficiently described by the spirit's complexion. Heat is spirit's defining characteristic, but the elemental heat of the terrestrial sphere is not adequate for its creative activities. In addition to the elemental heat, the spirit carries the "continually vivifying" celestial heat which it gathered from the heavens:

> The "instrument" of this power is twofold. One is celestial heat, which is continually vivifying, never completely used up, making animate creatures accept some conformity with the heavens. It works to bring together and sustain these things [animate beings] especially; that is the whole power of the heavens works to constitute and sustain the being of substances. This is the more worthy of the instruments . . . There is another secondary instrument, namely elemental heat.[49]

With the extraordinary complexion of this spirit, the informative virtue is able to properly organize substances.[50] As wonderful as this virtue is, even it needs appropriate material – sperm is the only material body with the potential to receive this celestial form.[51]

The informative virtue immediately goes to work on the material it finds in the womb, acting also as a nutritive soul to direct the growth of the organism:

**48** *Conciliator*, diff. 48 (f. 72r, col. 1, A–B): "Haec autem virtus divina praetacta prae aliis nobilitata generationis virtutibus, cum ipsa sit simplex et immaterialis fere. Non immediate materiam attingens, indiguit subiecto magis distincto eam vehiculante ad membrorum figurationes et formas dandas convenientes. Hoc autem fuit spiritus; est enim corpus subtile vehiculum virtutis. Eguit etiam instrumento, cum sit simplex sicut aliae mistorum virtutes."

**49** *Conciliator*, diff. 48 (f. 72r, col. 1, B): "Huius autem virtutis duplex existit instrumentum. Unum quidem est calidum coeleste, quod est vivificativum continue, nunquam vero peremptivum, faciens animata ad aliquam caeli conformitatem accedere. Agit enim ad consistentiam et continentiam eorum maxime, ceu omnis virtus coelestis facit constitui et in esse contineri substantias generatas. Et id dignius est instrumentorum . . . Est etiam ipsi aliud instrumentum secundarium, scilicit calidum elementare."

**50** See Hirai 2012: 24–30 for later criticism of Pietro's theory by Nicolò Leoniceno (1428–1524). Leoniceno judges that Pietro misreads Aristotle by saying that the heat of the formative virtue is celestial and "divine". Rather, Aristotle only meant to say that it is *analogous* to celestial heat, not that it *is* celestial.

**51** *Conciliator*, diff. 48 (f. 72r, col. 1, D): "In spermate est substantia prima potens recipere hanc virtutem, puta vitalem, et spiritus primus deferens hunc calorem est causa omnium partium spermatis, hunc autem dico coelestem."

> The matter in which the informative virtue operates with these two instruments [two kinds of heat] is the menstrual blood, drawn to the womb in the place of generation, which receives the due forms of the organs, and [the informative virtue] cares for it with nutrition and growth.[52]

To impose the proper structure on this mass of blood and semen, the informative power and its heat have four immediate tasks, presented in vivid detail by Pietro. The first is "to dissolve and extend the wetness of the semen with the power of its heat, so that it may become a receptacle of forms". Next, is "to strike and forcefully beat" throughout the whole humid mass, so that "many apertures and spaces are created, so that the heart and arteries may be constituted". Third, the informative power pauses to "inform" the parts of the body with their distinct identities and functions "because this is the function of spirit". Finally, the spirit "separates its own paths [i.e., arteries and nerves], from more humid paths" used by less subtle fluids, because the spirit "needs to irradiate quickly through the whole body, illuminating its home like a candle".[53]

After these four tasks, we are close to having an organic body suited to becoming endowed with a substantial form. At this point, the earlier powers in the informative virtue and the menstrual blood act together with the heat in the newly created heart to produce their own "flowing spirits", which "are not the cause of the formation of the heart, but rather are formed by it".[54] This new flowing spirit "confers vital power through the whole body and gives it the actual potential for living".[55] In other words, once the embryo has the organs to produce its own spirits, it has finally developed to the point that it is a natural body which potentially has life. Once the body is producing its own spirits, the informative virtue is corrupted and the body receives its single

52 *Conciliator*, diff. 48 (f. 72r, col. 1, C): "Materia vero, in qua virtus informativa operatur his instrumentis, sanguis est menstruus attractus a matrice in generationis locum, qui formas recipit membrorum debitas et ea demum nutricatione cum augmento custodit."
53 *Conciliator*, diff. 48 (f. 72r, col. 2, A): "Qui siquidem videtur quatuor actus circa principium generationis exercere. Quorum primus est dissolvere ac extendere humidum seminis virtute calidi sibi adiuncti, ut receptabile fiat formarum. Secundus est ferire et pulsare fortiter per totum humidum, quo plura causat foramina et concavitates constituuntur, ut cor et arteriae ipsius vasa constituantur. Tertius est informativam virtutem huius operis praesidem ad omnes membrorum informationes deferre; hoc namque dictum est spiritus officium esse. Quartus quoque est separare vias suas a viis humidi; non enim propter grossitiem ipsius et compactionem posset spiritus per vias illius velociter transire, cum oporteat ipsum celeriter totum corpus irradiare, ut illuminans domum velut candela."
54 *Conciliator*, diff. 48 (f. 72r, col. 2, B–C): "Quo peracto, vigore informativae cum aliis eius principiis ex materia menstruali, virtute prioris spiritus et calidi vehementer cordi inhaerentis, spiritus praetactus fluens generantur secundus per motum cordis et arteriarum . . . Sic igitur videtur iste spiritus duo facere in materia menstruali et differe a praedicto in eo, quod non est causa formationis cordis, sed potius formatur ab eo, corpore deinde iam organizato."
55 *Conciliator*, diff. 48 (f. 72r, col. 2, C): "Hic siquidem spiritus solum videtur virtutem vitalem deferre per corpus universum et dare illi potentiam vivendi actualem et nutriendi ampliorem."

new form: the rational soul, which takes on all the lower functions of the soul as well.[56] The whole process takes about 35 days for males and 45 days for females.[57]

Pietro's explanation for why the informative virtue cannot continue to exist as a formal principle is indicative of his interpretation of hylemorphism. He notes that not everyone shares his view that it must cease to exist. Because the informative virtue directs the early development of the embryo, some other natural philosophers say that it is the same as the soul.[58] Why should the same principle not continue to guide the later growth of the fetus? However, Pietro argues forcefully that the informative virtue and the soul must be different substances. The informative virtue cannot be the soul because the embryo has changed to the point that it is now a different entity. This necessitates a different substantial form: the soul, the form of a body with organs, as Aristotle maintained. The informative power is simply not this kind of form – one of its essential characteristics is that it has no organs:

> According to Aristotle, in some sense something divine can be apprehended in a thing existing separate from matter. [Informative virtue] is like this (as I said earlier), because it does not have some organ assigned to it as an instrument and therefore it is similar to the celestial powers. This could not be if it were to remain in the now organized body, for then it would be joined to an organ and it would simply be a natural virtue. For this reason, it is abolished when the body is formed.[59]

This clear distinction between the human soul and the informative virtue preserves the definition of soul as the actuality of a particular kind of organic body. At the same time, the corruption of the informative virtue also preserves the singularity of the soul as the body's only form.

---

56 *Conciliator*, diff. 48 (ff. 72r, col. 2, D–f. 72v, col. 1, E): "Dicendum quod virtus corrumpitur informativa, cum embryonem plasmaverit, sicut et nutritiva veterascens ex patre, ut dictum prius. Una nanque esset eius permanentia, postquam eius perfecerit actum, ut apparet etiam in augmentiva . . . Cuius quidem igitur corruptio erit, ut dicatur, circa 35 dies in mare, in foemellis vero circa 45 (et inde *differentia* 49); embrya enim circa haec formantur tempora. Quidam vero dixerunt ipsam corrumpi, quando anima introducitur intellectiva, ita ut omnes embryonis operationes sint ipsius."

57 Somewhat similar accounts of generation can be found in Dino del Garbo and Turisanus. See Siraisi 1981: 172–175.

58 *Conciliator*, diff. 48 (f. 72r, col. 2, B): "Cum igitur videamus principia generationis materialia maxime moveri transmutando se ad debitas formas animatorum et figuras recipiendas, oportet motorem adesse. Hic autem non potest esse nisi virtus informativa. Et ideo vocatur quandoque a physicis "anima". Sed quia huiusmodi virtus cum sit formalis et simplex per se non potens informare diversa, eguit instrumento, quod quidem diversificatum figurarum inducat diversitatem. Sic enim indiguit eam vehiculantem subiecto, quod et aliquo modo etiam rationem assumpsit instrumenti."

59 *Conciliator*, diff. 48 (ff. 72r, col. 2, D–72v, col. 1, E): "Quod ostenditur, quia ipsa, iuxta Aristotelem, est quid separabile existens a corpore in quibuscumque apprehenditur quid divinum. Hoc autem est, quia non habet (ut tactum est) membrum aliquod velut instrumentum sibi deputatum, et ideo virtutibus fuit assimilata coelestibus. Quod non esset, si corpore perduraret iam organizato. Tunc enim coniungeretur organo et virtus esset simpliciter naturalis. Propter quod aboletur, corpore formato.'

Using Pietro's discussion here in *differentia* 48, Bruno Nardi argues that Pietro's position on the soul is essentially identical with that of Thomas Aquinas.[60] Indeed, Pietro maintains the most distinctive Thomistic thesis of the unity of substantial form in man, arguing that all previous forms are corrupted, and their powers are subsumed within the new soul. Pietro also reinforces his account of the corruption of the informative virtue with arguments borrowed from Aquinas, arguing that, if previous forms did not cease to exist, there would be multiple actualities in the same person and the unity of the human being would be destroyed.[61] However, in the context of his entire response, he actually uses Aquinas in a rather limited way. For the vast majority of this mini-treatise on the animation of the embryo, Pietro has only been concerned first with the actions of the informative virtue, which acts as the nutritive soul in the early embryo, and the single human soul, which he barely mentions, but will be introduced at the end of the process.

Only once he has completed his own account does he restate arguments that can be attributed to Thomas. "Some people" say that there is first a nutritive soul and then a sensitive soul before the intellective soul supersedes them. He explains that, according to that account, the nutritive soul is corrupted and the body is reduced to prime matter for an instant before the sensitive soul comes about; in turn, the sensitive soul corrupts and the intellective soul supplies the body with its one form, containing all the powers of the two prior souls[62] Pietro then supplies an abridged version of a few common Thomistic arguments in support of this theory and against the idea of the plurality of forms. For example, he explains that the apparent discontinuity involved in the momentary reduction of the embryo to prime matter can be accounted for by general principles of generation and corruption and, at the same time, that the alternative doctrine of the plurality of forms is absurd because it would entail the existence of multiple entities, rather than a unified human being.[63] He concludes

---

**60** Nardi 1912/1958.

**61** Nardi 1912/1958: 5–7. Cf. *Conciliator*, diff. 48 (f. 72v, col. 1).

**62** *Conciliator*, diff. 48 (f. 72v, col. 1, E): "Quidam vero dixerunt ipsam [virtus informativam] corrumpi quando anima introducitur intellectiva, ita ut omnes embryonis operationes sint ipsius. Quod conantur ex Aristotelis 2 *De generatione animalium* persuadere. 'Propter quod communiores dixerunt, quod existente informativa, iam adest etiam nutritiva, sicut ostensum ex Aristotelem. Qua quidem recedente, et embryone usque ad materiam primam deducto, sensitiva anima priori supervenit corrupta, virtutem habens vegitandi et nutriendi. Hac denique ut priori ad materiam deducta primam propter alterationes ad hoc praecedentes anima introducitur priori abolita intellectiva fulcita." Cf. Amerini 2009.

**63** *Conciliator*, diff. 48 (f. 72v, col. 1, H): "Ad aliud dicendum, quod quia id evenit in instanti, non percipitur, cum mutatione introducitur et non motu (*Physica*, 5). Quod et sentire apparet Aristoteles, volens (*De generatione et corruptione*, ipsam fieri, cum hoc totum transmutatur in hoc totum, quod non esset, nisi ad materiam primam pervenientur. Alioquin siquidem non esset generatio, verum potious alteratio subiecto permanente. Adhuc, nisi prima corrumperetur forma, altera enti eveniret existenti in actu. Quod autem sic occurit, est accidens. Similiter et cum esse sit a forma substantiali, habenti

by noting that this theory, briefly summarized at the end of this *differentia*, can be found more completely elsewhere.[64] He does not specify where, but he very well could have Aquinas in mind. All of this is a supplement to Pietro's much lengthier treatment of the material formation of the organism, but he clearly regards Aquinas' arguments as being in basic accord with his conviction in the essential unity of body and soul and his emphasis on the bodily component of human nature.

However, if Pietro's position is identical with Aquinas', this is not the way the Dominicans at St. Jacques (Jacobites) saw it when they accused him of materialism. If we assume that the Jacobites adhered to Thomas' teachings (or at least some version of them), they obviously did not appreciate Pietro's use of this doctrine here. It is not difficult to imagine why. It is possible that the Jacobites had a particular reading of Thomas that insisted on a specific interpretation of his doctrines. It is also probable that, as the Dominican order defended the thesis of the unicity of substantial form, they wanted to distance this thesis from its use by a physician who defended Galenic psychology and seemed to veer toward materialism.

This accusation by the Jacobites reminds us that the true focus in this *differentia* and elsewhere is not the soon-to-be-introduced rational soul, but the body, with the spirits and informative virtue that organize it. It is impossible to read far into his discussion without getting the sense that Pietro is far more impressed by the informative virtue than he is with the soul itself. He repeatedly interrupts his narrative of embryonic development with encomia to the divine and celestial powers borne by the informative virtue.[65] Galen, he says, calls it the "highest art, the ruler, an intellect or mind".[66] Pietro identifies it with the "intellectus vocatus" of Aristotle's *De generatione animalium*, II.3, which enters the body from without and was typically understood as the rational soul itself.[67] It is such a marvelous thing because it defies the physics of the terrestrial world, organizing the embryo in a way that would otherwise not be possible. The soul actually seems to be more explicable than the informative virtue; at least the soul has an appropriate body, while the informative virtue does not.

---

quidem plures formas, et plura esse haberet, ut et nihil hic vere unum existeret." On Aquinas' development of these arguments, see Pasnau 2002: 120–130.

**64** *Conciliator*, diff. 48 (f. 72v, col. 2, E): "Nosce quoque, quod plura in hac differentia collecta sparsim in aliis conscripta."

**65** For the influence of the heavens on the human body, see Jacquart, 1993, and Vescovini 1997.

**66** *Conciliator*, diff. 48 (f. 71v, col. 2, G): "Rector autem huius tam divini operis virtus est dicta informativa ab anima parentis decisa in actum per impulsionem coeuntis incitata, quam Galenus, *De naturalibus* secundo, appellat "summam artem", "praesidem", et "intellectivam" sive "mentem" (sine mente *ed.*).

**67** *Conciliator*, diff. 48 (f. 71v, col. 2, G): "Aristoteles autem appellat 'intellectum vocatum' sive 'intellectivam divinam' . . . Nominavit autem eam Aristoteles 'intellectum vocatum' ad differentiam intellectus potentionalis et agentis pars existentium animae intellectivae, ut tertio *De anima* inquit." For further discussion of the *intellectus vocatus*, see below, Chapter 4: 89–91.

The arrival of the soul itself – which is after all the topic of this *differentia* – is treated almost in passing as the predictable end of the accomplishments of the informative power.[68]

We have to turn to *differentia* 71 for a more complete account of the arrival of the soul. But even there, as Pietro celebrates the gift of the soul from God, he is most attentive to the marvelous body. In a long passage, somewhat confused because it incorporates passages from so many authorities, Pietro extols the body able to receive the soul:

> When the mixture of the elements is more beautiful and more perfectly equal, so that nothing more subtle and beautiful can be found, as in the sperm of man, then it becomes apt to receive from the giver of forms, the most beautiful form of [all] forms, which is the human soul.[69]

Such passages seem to push the limits of physiological explanation for the introduction of soul, even if the informative virtue is itself "divine" and that ultimately the soul comes directly from God. However, it remains that the beautiful complexion and disposition of the body created by the informative virtue coincides with the status of the soul in the strongest manner implied by hylemorphism. The relation of the informative virtue to the initial combination of blood and sperm does not seem so strong. Even if Pietro's doctrine is technically orthodox, it is not difficult to see why the Jacobites did not approve of his emphasis.

If James Bono is correct that Albertus and other philosophers sought to clearly distinguish between the material nature of spirits and the immaterial soul, the opposite seems to be true in Pietro's thought. Instead, Pietro seems to return to the twelfth-century Neoplatonic idea of spirits in emphasizing that their nature transcends matter and is "almost" incorporeal. Together with their two-fold complexion (celestial and elemental heat) they unite with the formative virtue to prepare for the immaterial form of the human soul. Pietro's efforts to define all this show the limits of his vocabulary to do so: he repeatedly emends his owns definitions, seeking precise terms in which to describe the mediating role between soul and body, seeking to keep the spirits as immaterial as possible, as they mediate between opposed substances.

---

**68** The introduction of the soul is listed among the four tasks of the new *spiritus fluens*. *Conciliator*, diff. 48 (f. 72r, col. 2, D).

**69** *Conciliator*, diff. 71 (f. 108v, col. 1, G): "Quando mistio elementorum fuerit pulchrioris et perfectioris aequalitatis, qua nihil possit inveniri subtilius et pulchrius, sicut est sperma hominis, tunc fiet apta ad recipiendum a datore formarum formam formis pulchriorem, quae anima extat humana." Note that this passage does not seem to be entirely consistent with *differentia* 48. Here he indicates that the soul enters when the elemental complexion is apt but does not mention organs.

# Chapter 4
# The Organic Soul and the Intellect

Pietro shared a general idea of the organic soul and its faculties with his contemporaries.[1] Besides being the first actuality of the body, soul is also the organizing principle for all the faculties, powers, or capacities situated in the organs (hence "organic") of any living creature. Theologians and natural philosophers, following Aristotle and the later contributions of Avicenna and Averroes, conceived these faculties as a hierarchy ranging from the simplest vegetative activities necessary to the maintenance of life, to perception and motion, to memory and rational thought. In the century before Pietro's career, scholars had drawn on traditions in philosophy and medicine to develop a general consensus about the localization of these faculties in the body and the physiology that corresponds to their activities.

At the top of the hierarchy was the rational soul with the intellect, the part of the soul specific to humans. Pietro describes the parts of the intellect at least three times in his work, in addition to his other thoughts about the human soul scattered through his texts. In his descriptions, Pietro reveals no great interest in metaphysical speculations about the essence of the intellective soul, but only in naming its perceptible faculties.[2] Without feeling the need to elaborate, he takes for granted the almost universal agreement that the intellect is separable and immaterial. However, his thoughts on the intellect as a whole all tend in a single direction: to describe the necessary material conditions for the intellect to function that will grant him the largest possible latitude to talk about intellect in terms of his medical philosophy. For this, he emphasizes the notion that the imagination and the informative virtue (or *intellectus vocatus*), which depend on complexion and spirits like every other power of the soul, are necessary for the intellective soul.

## The Organic Soul

As the guiding idea of conciliation suggests, Pietro's description of the soul's faculties (*virtutes*) contained in the *Conciliator* is notable for its inclusiveness. In *differentia* 57, which asks whether the vital *virtus* is distinct from the other powers, Pietro combines multiple definitions of *virtus* into a single, somewhat unusual hierarchy, held together loosely by the structure of the Aristotelian tripartite soul and its vegetative, sensitive, and rational parts, but especially by Avicenna's elaboration of this structure.[3] Pietro

1 For an overview of the conception of the organic soul shared by most scholars in Pietro's time, as well as some of the topics of disagreement, see Hasse 2010. For the continued influence of the idea of the organic soul formed around 1300, see Park 1988.

2 According to Zupko 1997, there was a growing empirical tendency in the study of the soul in the late thirteenth and fourteenth centuries to focus on perceptible powers rather than the soul's essence.

3 *Conciliator*, diff. 57 (ff. 83v–85v): "Utrum virtus vitalis sit altera, necne?"

https://doi.org/10.1515/9783110612639-005

understands *virtus* in a very broad manner: it can simply convey possibility; it is the efficient cause of an action in the straightforward way that no action can happen without there first being the potential for it to happen. That potential, however immediate or distant it may be, is a *virtus*. Thus, *virtus* can be rendered in English as a *faculty, potential, power, function,* or *capacity for different types of action*. Pietro gives four ways he has found *virtus* to be used by Aristotle:

1) it can be a material quality, such as the heat possessed by fire;
2) it can be the highest potential of a creature, such as the capacity of a newborn to acquire rational thought;
3) it can be a fully developed habit, in the sense of a moral virtue;
4) virtues can be used to name the powers of the tripartite soul.[4]

Pietro adds that the *medici* have a different method of classifying the virtues, which he will need to consider as well. However, according to Pietro, the medical virtues do not conflict with the more elaborate philosophical division but complement it by offering another perspective.

In this *differentia,* he pulls these various definitions into a single "chain of the virtues" (*catena virtutum*) encompassing all the virtues of the body and soul. He presents this chain in a straightforward manner, without the polemical tone he sometimes adopts, but there are characteristic features of this account that reveal distinct priorities in his thought. Notably, the inclusion of the elemental qualities (already mentioned at the beginning of Chapter 2) and the moral virtues in the same overall account has the effect of flattening the difference between definitions of *virtus* that otherwise occupy different places in the Aristotelian corpus. His account progresses from the lowest to the highest capacities of the body and soul, suggesting to the reader that somehow all of this shares a common explanatory system. Throughout, Pietro reminds his readers that all of these virtues are "perfections of the organic body".[5]

4 *Conciliator*, diff. 57 (f. 83v, col. 1, E–F): "Propter primum vero sciendum, quod virtus est causa quaedam efficiens (*Pharmacorum*, 1) et omnis operis initium (prima primi [*Canonis*]). Ipsam quoque in presentiarum quadrupliciter reperio ab Aristotele suscipi.
1) Uno quidem modo pro qualitate, cum ab ipsa omnis procedat operatio in natura (iuxta illud, [vide] primi prima). Cumque elementa agunt et patiuntur suis virtutibus.
2) Aliter quidem pro ultimo aliquius posse (unde primo *[De] caelo et mundo*). Virtus est excellentie, quod est ultimum potentie.
3) Adhuc pro virtute morali, que est "habitus electivus in medio quo ad nos, existens determinate ratione, ut utique sapiens determinabit" (*Ethicorum*, 2).
4) Sumitur et pro potentia anime secundum quod eam Aristoteles dividit in vegetativam, sensitivam, et rationalem."

5 E.g. *Conciliator*, diff. 57 (f. 83v, col. 2, F): "[V]irtus rationalis, vel humanus . . . est perfectio praetacti organici corporis . . ."

As was customary in medieval accounts of the soul's faculties, Pietro arranges them hierarchically based, (1) on the subtlety of the organic material used by the soul, and (2) by the level of abstraction from the particular sense object in the sensory and rational parts of the soul.[6] See the table below p. 78 for the full chain of virtues from this *differentia*. His subsequent division is comprehensive, but abbreviated, and is sometimes difficult to follow without reference to Avicenna's *De anima*.

Pietro begins with the vegetative, or natural, soul. The vegetative powers include nutrition, growth, and generation – the powers necessary to produce and sustain life. Generation is divided into four parts: one for the existence of the human species, and three for the production of new individuals and preparing for the divine informative virtue, on which he refers the reader to the greater detail in *differentia* 48 discussed in the previous chapter. The augmentative powers are divided into two, one incorporating and the other assimilating (or informing). The nutritive powers are attractive, digestive, retentive and expulsive. These nutritive faculties are more passive than the other two parts of the vegetative soul (which do more shaping or informing of matter) and can be almost entirely accounted for by the four prime qualities, which serve them.[7]

The sensitive soul is divided into the powers of movement and apprehension. He further subdivides each power, generally following Avicenna's faculty psychology.[8] The motive power has an appetitive faculty, which commands motion, and an effective faculty, which actually produces the motion.[9] The command of the appetitive power takes the form of a desire to attain something (concupiscible power) or to repel something (irascible power). These desires are physically manifest in emotions shared by the body and soul and spread through the organism by the heart and blood. The

6 Which of these two hierarchies was most emphasized depended on the scholar. Albertus Magnus had a much greater interest in the material hierarchy than Thomas Aquinas, for example. On this, see Mahoney 1982: 602–622.

7 *Conciliator*, diff. 57 (f. 83v, col. 1, H): "His vero quatuor virtutibus immediate demumque aliis cunctis tanquam instrumenta inaequaliter quatuor deserviunt qualitates primae: caliditas videlicet, frigiditas, siccitas et humiditas." The brief account of the vegetative powers in Avicenna, *De anima*, 1.5 (ed. Van Riet 1972: 101–102), corresponds to Pietro's and includes the idea of the four qualities "serving" the nutritive powers.

8 *Conciliator*, diff. 57 (f. 83v, col. 1, H): "Virtus autem sensitive, seu anima, est perfectio organici corporis secundum quod apprehendit et movetur appetite, duplex existens: motiva scilicet at comprehensiva." Extensive use of Avicenna's *De anima* preceded the *De anima* of Aristotle. As a result, most later readers read Aristotle with Avicenna's faculty psychology already in mind. On the reception of Avicenna's psychology see Hasse 2000.

9 *Conciliator*, diff. 57 (f. 83v, col. 1, H–col. 2, E): "[N]am una movet, eo quod imperat motui, vel quia efficit motum. Prima quidem est appetitiva (seu desiderativa) cordi ([diff.] 38) insita, duplex etiam existens. Una quidem vis concupiscibilis, quae imperat moveri, ut appetitivi adiiciantur congrua; reliqua irascibilis dicta, imperans moveri, repellantur nociva. Vis autem motiva motum efficiens, nervis et musculis est infusa, quae chordas et ligamenta contrahit et itidem relaxit propter quod moventur membra et iuncturae dilitatione et fixione." See Avicenna, *De anima*, 1.5 (ed. Van Riet 1972: 82–83).

**Table 1:** *Conciliator*, diff. 57: *Catena de viribus animae*: Chain of the Soul's Faculties (with the running total number of faculties).

- 0) Four Qualities (4)
- 1) Vegetative Faculties
  - a) Four Nutritive faculties (8)
  - b) Two Augmentative faculties (10)
  - c) Four Generative faculties (14)
- 2) Sensitive (or Animal) Faculties
  - a) Motive
    - i) Appetitive or Commanding
      - (1) Concupiscible (Desiring) (15)
      - (2) Irascible (Repelling) (16)
    - ii) Effective (Estimative) (17)
  - b) Comprehending
    - i) Manifest Senses
      - (1) Touch (18)
      - (2) Taste (19)
      - (3) Smell (20)
      - (4) Hearing (21)
      - (5) Sight (22)
    - ii) Internal (or Occult) Senses
      - (1) Concerning Sensible Forms
        - (a) Common Sense (23)
        - (b) Imagination or Phantasia (24)
        - (c) Cogitation in Humans/Imaginative in Animals (25)
      - (2) Concerning Intentions
        - (a) Virtues abstracted from form, with some attachment to matter
          - (i) Estimation (26)
          - (ii) Conservative (27)
          - (iii) Memory (28)
          - (iv) Reminiscence = Rational faculty (29)
        - (b) Virtues abstracted from form, with no attachment to matter = Rational Faculties
- 3) Rational Faculties
  - a) Active
    - i) Mechanical Arts (30)
    - ii) Prudence, including ethics, economics, and politics (31)
  - b) Speculative
    - (1) Potential Intellect
      - (i) Material (32)
      - (ii) Potential (33)
      - (iii) Perfected Intellect (34)
    - (2) Agent Intellect (perhaps 35)

effective power, infused throughout the nerves and muscles, actually produces movement after either assenting to the desire of the appetite or obeying an instinct which bypasses the appetite altogether.[10]

Pietro divides the apprehensive part of the sensitive soul into two levels of sensation, the manifest and the occult, or hidden. The "manifest" senses (usually called the "external" senses in the Avicennan taxonomy, because they involve exterior organs) are the five senses of sight, hearing, smell, taste, and touch.[11] It is the job of these five manifest senses to perceive the four primary qualities (i.e. hot, cold, wet, and dry) and the secondary qualities derived from them (e.g. hard, soft, bitter, sweet, etc.). For the "internal" or "occult" senses, so-named because they are located within the body, he names seven separate faculties, which is a rather unusual number.[12] Avicenna named five internal senses, while Averroes identified only four, and most people followed one of them in this.[13] Arguments about whether there are four or five internal senses typically focused on whether the estimation really exists as a distinct faculty, as Avicenna claimed and Averroes denied. Pietro follows Avicenna and adds the extra faculties by subdividing the memory, but does not signal that this is a contentious issue in any way. Three of the internal senses deal with what Pietro calls "simple form", that is, they directly apprehend the sense perceptions of external sense objects collected by the five manifest senses.[14] These are the common sense, the imagination, and cogitation, which are respectively powers of perceiving, storing, and composing the external sense perceptions. The common sense receives the species from the external senses, which it then sorts and compares. In doing this, it perceives "common sensibles" like motion, shape, and extension that are not exclusive to any single sense. The imagination stores the simple forms (or images) from the common sense. The cogitative (confusingly called the imaginative in non-human animals) combines and divides the images stored in the imagination, creating new images in infinite configurations (more or less corresponding to the modern use of the word "imagination").

10 See Avicenna, *De anima*, 4.4 (ed. Van Riet 1968: 57–58).

11 *Conciliator*, diff. 57 (f. 83v, col. 2, E): "Quae autem comprehensiva duplex existit: 1) quaedam quidem manifesta, quae in quinque dividitur sensus, ut visum, auditum, odoratum, gustum, et tactum."

12 On the internal senses, see Wolfson 1935, Harvey 1975, and Tachau 1993. The interaction of the internal senses in producing health is explained in the next chapter.

13 Hasse 2010. Roger Bacon and Albertus follow Avicenna, while Aquinas follows Averroes. For Avicenna's five internal senses, see Avicenna, *De anima*, 1.5 (ed. Van Riet 1972: 85–90). Averroes names only the common sense, imagination, cogitation, and memory. See, Averroes, *Commentarium magnum in Aristotelis De anima* (ed. Crawford 1953: 416).

14 *Conciliator*, diff. 57 (f. 83v, col. 2, E): "Altera occulta et illa bina, quia vel negotiatur circa formam, vel intentionem. Et si primum, vel operatur circa formam simplicem, et hoc est triplicitur, aut eam comprehendendo, et ita sensus communis vel conservendo et retinendo, et tunc imaginatio vel phantasia, que secundum Algazelem est formarum archa, aut componendo, componit enim et dividit eam, et ita imaginativa virtus brutalis, seu cogitativa rationalis."

The other four occult senses function at a higher level of abstraction. They apprehend "intentions", the quasi-material entities that contain information imperceptible to the external senses, such as hostility and friendship.[15] The estimative power is still "partially connected to matter", by which Pietro means that it abstracts intentions directly from the sensory forms. Three types of retentive powers store these intentions: the conservative power, the memory, and reminiscence. These are distinguished according to how long they store the information and what kind of information they contain. This threefold memory is highly unusal and breaks with Avicenna, who named only a singular memory and rejected the idea of an intellectual memory.[16] It is probably based on Aristotle's *De memoria et reminiscentia* in the *Parva naturalia*, where Aristotle distinguishes between levels of memory.[17] With reminiscence begins the transition to the rational powers because it stores abstract concepts used in creating syllogisms and is possessed only by humans.

In the description of the rational powers that follows, Pietro characteristically deemphasizes the distinction between the powers of the sensitive soul and those of the rational soul. Where most philosophers would have called attention to a sharp distinction, Pietro classifies the rational powers within his group of occult virtues. Thus, his category of "occult virtue" contains both sensitive and rational powers, which has the effect of blurring the difference between these parts of the soul. In the chain of virtues under consideration here, Pietro inserts the rational virtues into the hierarchy of occult powers as those that regard intentions now completely separated from their matter. Reason has active and speculative parts. The active virtues include the seven mechanical arts and the moral virtues, which Pietro classifies as subcategories of prudence. The inclusion of the arts and the moral virtues here is at the very least atypical and an example of Pietro's inclusiveness. However, if we keep in mind the view of Aristotle's *De anima* as an over-arching framework for psychological issues, and many of his other books as elaborations of its parts (as the *Parva naturalia* and books on animals were certainly considered), there is no reason that ethics cannot be regarded as the amplification of the virtue of active reasoning. This seems to be the motivation for Pietro's placement of moral virtues here. These active virtues plan particular actions after the consideration of universal principles.[18] Finally the highest part of soul, the speculative

15 *Conciliator*, diff. 57 (f. 83v, col. 2, E): "Siquidem circa intentionem seipsam exercet, aut hanc abstrahendo a forma cum appenditiis tamen materie, ut hic et nunc, seu absque, sed universaliter. Quod si primum, duplicititer: aut ipsam denudando, particulariter tamen a forma, et tunc estimativa; vel retinendo, et hoc tripliciter: aut sine interruptione aliqua ipsius, et temporis quasi distantia, et hec dicitur conservativa; vel contra, et sic reminiscitiva, que velut quis extat syllogismus; aut mediocri, et tunc memoria, que communior est illa."

16 Black 2010: 325–326.

17 Aristotle, *Parva Naturalia*, ed. and transl. Hett 1936.

18 *Conciliator*, diff. 57 (f. 83r, col. 2, F): "[A]ltera activa, quae principium est movens ad actiones cogitationis singulas." All of the rational virtues, active or speculative, function "cum deliberatione vel meditatione universalia suscipiens." This distinguishes the active reason from the "thought" that goes on in the sensitive soul, which only uses particulars, not universals, in its reasoning.

virtue, is divided into potential and active parts, which can be variously subdivided depending on the authority one follows. Unusually, there is no place in this hierarchy for a will, or rational appetite. Pietro's thoughts on the intellectual soul are explained more completely below. To this tally, Pietro appends the four primary qualities (hot, cold, wet, and dry) at the bottom of the hierarchy of the soul's virtues.

With the qualities, the exhaustive list totals 34 or 35 powers (depending, he says, on whether the agent intellect is considered part of the individual soul).[19] Pietro is clear that these individual capacities merely name the observable powers of the singular soul; they are not separable and substantial "parts" of the soul.[20] They are only parts in the way that accidents are "parts" of a substance, so that each power of the soul ultimately refers to the complete and singular soul as substantial form of the body. The intellect, as the highest power, ought to be the sole master; the other powers are its "servants" (*famulans* or *annules*).[21] The four qualities ought to be servants to every other faculty, just as all the inferior powers should serve their superiors. This is a description of all the virtues possible in a perfect animal.[22]

Pietro next turns to the medical description of the virtues; he notes that, as an *artifex*, a *medicus* has more particular (*singularior*) concerns than a philosopher, viewing the virtues as the three distinct systems manifestly connected to the three primary organs (or four if the testicles are included for generation).[23] By this he means that the medical tradition is less inclined to impose a strict hierarchy of value on the soul, because each system is equally important for the preservation of health. According to physicians, he says, the virtues are divided into the natural, vital, and animal, which are separate systems originating respectively in the liver (with the testicles), the heart, and the brain. The natural powers are those of nutrition and growth; they

19 *Conciliator*, diff. 57 (f. 84r, col. 1, B): "Ex his itaque diligenter computatis, apparet quod virium animae catena 34 aut 35 cum eisdem deservientibus constituitur annulis." See below for Pietro's ambivalence about the number of faculties in the intellect.

20 *Conciliator*, diff. 57 (f. 84v, col. 1, G): "Et itidem nosce quod divisio animae in suas potentias est divisio subiecti in sua accidentia, ceu pomi in colorem, saporem et odorem, et non generis in species (Commentator, 1 *De an.*)." Also see diff. 38 (ff. 58v, col. 2, H–59r, col. 1, A), where Pietro argues that there must be a single ruling principle in the body, just as a good city has one ruler: "Existimandum animal constare velut civitatem bene legibus rectam. In hac vero unus praeest, in quem omnia tendunt tandem, ut decet, civilia."

21 *Conciliator*, diff. 57 (f. 84r, col. 1, B): "Inter quos annulos, intellectus in actu, vel accomadatus seu acquisitus, principans extat tantum et servitus omnifariam. Qualitates vero quatuor primae famulantes, et dominatae omnino, quae vero in medio se mediocriter habent."

22 *Conciliator*, diff. 57 (f. 84v, col. 1, G): "Amplius sciendum quod quaesitum potest referri aut ad totum genus hic viventium, vel partem, sicut animal et praecipue perfectum."

23 *Conciliator*, diff. 57 (f. 84r, col. 1, B–C): "Medicus autem artifex singularior virtutem dividit in naturalem, vitalem, et animalem, cum tria vel quatuor manifestentur membra, velut principia cor, cerebrum, et hepar, cum testiculis."

process nutrients and restore the body.[24] They do this by means of the blood and natural spirits distributed throughout the body through the system of veins originating at the liver. The reproductive powers are distinct in function and location because they maintain the existence of the species instead of the individual and are located in the testicles, although this was a topic of dispute among authorities.[25]

Pietro says that the physicians locate the center of animal powers in the brain, which distributes its spirits through the nerves.[26] These spirits enable acts of sensation and voluntary acts of motion and discretion. In addition to the brain, the organs of the animal virtue include all the other organs involved in sensation such as the eyes, nose, tongue, nerves, muscles and diaphragm. Animal virtue most resembles action "purely of the soul", since it is the "principle of comprehending and moving arising from voluntary comprehension".[27] In other words, its actions are freer because its organs are less material and therefore less governed by material necessity. In the functions of the animal virtues especially, we do not always act in correspondence with the rules of nature.

The natural and animal virtues fit rather easily with the philosophical account, generally corresponding to the natural powers and to the powers of sensation. The situation was different with the vital virtues, which seem to transcend all other powers. The vital virtues are seated in the heart and are distributed in the arteries by vital spirits. Its tasks overlap with both natural and sensitive functions. Its first job is to maintain life by distributing blood and heat through the body, involuntary activities that are similar to the natural part of the soul. The heart and its vital spirits have a "passive" role as the seat of the affections of the soul, such as anger, sadness, joy, and fear. In this role it "seems to be almost the same as those [powers] called desiderative, including concupiscible and irascible" from the philosophers' account of the sensitive soul.[28] But the vital virtue also precedes all other powers as "the first root of all virtues", supplying the basis for all life and preparing the parts of the body for receiving the sensitive and motive powers and all other life activities, as described in

---

24 *Conciliator*, diff. 38 (f. 59r, col. 1, D): "Membra quoque naturalia sunt duplicia: quaedam, ut alimentum per ipsa altera tum convertatur in corpus, ut restauretur quod deperditum in ipso, ad hoc ut permaneant individuum. Sunt autem haec: os, dentes, oesophagus, venter, intestina, hepar, fel, splen, renes, vesica, glandulasecus eam, et venae. Alia vero ut genitalia . . ."

25 *Conciliator*, diff. 38 (f. 59r, col. 2, A). Cf. diff. 34, "Utrum sperma a toto corpore, sive ab omnibus decidatur membris, necne", where Pietro defends the testicles as the organ responsible for the production of sperm.

26 *Conciliator*, diff 38 (f. 59r, col. 1, C).

27 *Conciliator*, diff. 57 (f. 84v, col. 1, F): "Similiter, virtus animalis potest differenter sumi, vel prout dicitur ab anima secundum quod actus, vel principium extat comprehendendi et movendi, ex aliqua proveniens voluntaria comprehensione."

28 *Conciliator*, diff. 57 (f. 84r, col. 1, C): "Alia vero patiens, sive ad haec operata, a qua passiones seu accidentia proveniunt animae – ut ira, timor, gaudium, et tristitia, et quae talium. Et ista fere videntur desiderativa dicta, sub qua irascibilis et concupiscibilis continentur."

the previous chapter.[29] For this reason, most of Pietro's contemporaries in medicine maintained that vital virtue should not be identified with any part of the tripartite Aristotelian soul.

## The Intellect and its Parts

As a rule, medieval scholars shared the assumptions that the intellect was immaterial and could subsist independently of the body.[30] The premise that the intellect is free from matter – and free from the laws that govern material things – was considered essential to safe-guard free will and the possibility for the most abstract thought. In the thirteenth century, philosophers had devoted considerable energy to prove that these principles could continue to be maintained even as they incorporated Aristotelian hylomorphic ideas about the soul into their doctrines.[31] The potential and agent intellect, which in several passages Aristotle suggested possess a different nature from soul conceived as form of the body, and thus might be somehow separate from the material of the body, offered a promising means of reconciling the Aristotelian tradition to these precepts.[32] Without contradicting the idea that some part of the intellect must be immaterial, Pietro took the opposite approach to most of his contemporaries. Rather than looking for evidence that the intellect is immaterial, he strove to define the largest possible material realm in which to discuss the highest rational functions in the intellect.[33]

There has probably been more attention paid to Pietro's thoughts on the intellect than they deserve. This interest can be attributed to the fact that when Pietro had trouble with the inquisition, among the 55 doctrines he tells us were cited by his Dominican accusers, the only one we know for certain concerns the intellect, specifically its origin: that he maintained that the intellective soul is brought forth from the potency of matter (see Chapter 3: 67–73).[34] Even if we believe Pietro's protests of innocence, the suspicion provoked by his thoughts on the subject suggests that his ideas did not appear wholly conventional.

---

**29** *Conciliator*, diff. 57 (f. 84v, col. 1, C): "Virtus vitalis est vis, quae cum in membris recepta fuerut, praeparabit ea ad recipiendum virtutem sensus et motus, et operationes vitae, quibus etiam timoris et irae adminiscentur motiones."

**30** For some of the shared ideas about the intellect, see Pasnau 2012b.

**31** For a brief summary of various strategies of reconciling Aristotelian ideas to orthodoxy about the soul, see Kuksewicz 1982, or Dales 1995 for a longer study of issues concerning the rational soul in the thirteenth century.

**32** Notably, Aristotle, *De anima* 3.4 = 429a10–429b10 (ed. and transl. Hett 1936: 135–137).

**33** By the early fourteenth century, several philosophers questioned whether the immaterialty of the soul could be proven through natural philosophy, rather than simply taken on faith. See Pasnau 2012b: 492–498.

**34** See Introduction, note 12, and *Conciliator*, diff. 3, ad. 4.4 (f. 8r, col. 1, A–B).

This interest is also due to a famous (mis)reading of Pietro's ideas. According to Ernst Renan's influential work on Averroism, Pietro not only held the Averroist doctrines that there is a single potential intellect shared by all individuals and that the imagination is the highest part of the soul specific to each person, (thus simultaneously making him an Averroist with regard to the intellect and a materialist with regard to an individual person), but that his thought was instrumental in creating the notorious school of Paduan Averroism.[35] Renan's interpretation was expanded in Ferrari's biography of Pietro d'Abano in 1900 and repeated by other influential readers.[36] In his study of Pietro's remarks of the generation of the soul, Bruno Nardi noticed that if Pietro's self-reported charge that the intellective soul is brought forth from matter is accurate, this is not an Averroist doctrine at all, but instead would suggest a materialist position associated with Alexander of Aphrodisias or Galen.[37] Nardi concludes that Pietro was certainly not an Averroist; nor was he a materialist. Instead, as we saw in Chapter 3, he determines that Pietro holds that the intellectual soul is divinely infused into material that has reached a proper state of development, where it develops as the single substantial form of an individual human.[38] Thus, Nardi finds Pietro's position to be rather conventional and more or less the same as Thomas Aquinas.[39] Dag Hasse looks at the description of the intellect in the chain of the soul's faculties from *differentia* 57, and reports that Pietro adopts an Avicennan doctrine of four intellects, with a three-part potential intellect belonging to the individual and a shared agent intellect.[40] Hasse further notes that Pietro associates the three potential intellects with Aristotle's single potential intellect, a reading that had been common since the first half of the thirteenth-century and certainly not philosophically controversial. Graziella Vescovini suggests, without citing any specific text, that Pietro adheres to a rather materialist position similar to Alexander of Aphrodisias or Galen.[41] More recently, Aurélien Robert, while not discussing the intellect in particular, argues that Pietro follows Aristotle and Averroes in the primacy of form over matter in the generation of substances, and is therefore not a materialist if materialism is taken to mean a doctrine by which there are no substances other than matter.[42] He also sensibly argues that assigning labels to Pietro's theory on the soul is not very useful.

Pietro describes the parts of the intellect three times in his work: once in the chain of the soul's virtues from *differentia* 57 discussed above, and twice in his *Exposi-*

35 Renan 1852: 258–259.

36 Ferrari 1900: 347–353.

37 Nardi 1912/1958: 2–3.

38 Nardi 1912/1958: 12. Pietro's defense of the unity of substantial form in the human organism, a position defended by Aquinas, is clear in differentia 48, discussed in the previous chapter.

39 Of course, Aquinas' ideas were hardly above suspicion in the period after 1277. See Wippel 1995 and Thijssen 1997.

40 Hasse 2000: 650–652.

41 Vescovini 2006: 57–58.

42 Robert 2017.

*tio Problematum*. The first thing that is apparent is that his descriptions are not consistent.[43] This inconsistency can be accounted for by the fact that Pietro never intends to propose a definitive doctrine on the intellect. Instead, his descriptions occur in the course of a commentary, in response to an objection, or another circumstance that causes them to vary depending on the needs of the exposition.

In *differentia* 57, he itemizes the parts of the intellective soul, just as he does for the other parts of the soul. After describing the practical parts of the intellect, which includes the moral and political virtues, he then turns to the speculative intellect, which he divides into potential and active parts. The potential intellect has three levels that correspond to different degrees of perfection:

1) The material [intellect], which has no form, but is the subject of all [form], as its first potential;
2) Another [intellect], like second potential, is when, in the material potency, *per se nota* principles are applied to intelligible species. From these [acts], [the intellect] is led to second intelligibles; and from these, to known principles. For these first principles are pre-existing propositions, coming *per se* to a person who has [an intellect]: such as the "being" or "not being" of something, or that every whole is greater than its part. This intellect is called "potential", namely the potency of [the intellect];
3) The third is called the perfected intellect, which is able to actually think whenever it wishes.[44]

Notice that this is not a list of distinct "parts" of the intellect which exist simultaneously, but stages of the development of the intellect. As Hasse points out, Pietro basically follows the conception of a three-fold potential intellect and a single active intellect, proposed by Avicenna in *De Anima* I.5.[45] The "material" intellect (which is not literally "material", but merely formless), is a distant first potential, like the potential for a baby to write. The second level of potential is when *per se nota* propositions that require no proof, such as the proposition that every whole is larger than a part, are used to analyze the "intelligible species" that have been prepared for the intellect in the sensitive soul through the process of abstraction from sense perceptions. The third potential is when the intellect is perfected and able to actually think whenever

---

**43** Hasse also observes that the texts are not identical and that an examination of them could alter his conclusions.

**44** *Conciliator*, diff. 57 (f. 57, 4G–H).: "Quibus quidem potentiis triplex proportionatur intellectus: [1] ut materialis nullam habens formam, sed subiectum existens omnis, ut ipsius potentia prima. [2] Est et alius relatus potentiae secundae, ut quando in potentia materiali habentur de intelligibilibus per se nota, ex quibus acceditur ad intelligibilia secunda ex eis nota principia. Prima namque sunt propositiones priores per se ad habentem venientes (4 *Meta.*), ceu de quolibet esse, aut non esse, ac omne totum maius sua parte. Et hic intellectus est potentialis dictus, ut eius potentia. [3] Tertius quoque est dictus perfectionis intellectus, qui potest actu quando voluerit intelligere."

**45** Hasse 2001: 650–652.

it wishes. This emphasis on development made Avicenna's version of the potential intellect popular with medical authors.[46]

The fourth virtue of the intellect is the active part:

4) However, when the intellect actually understands itself to be understanding, it is called *"in effectu"* and then the intellect called *"accomodatus"* is connected and united to it from outside [the body] (Avicenna, *De anima* 1). Or it is acquired from an intelligence, which he [i.e. Avicenna] says is the agent [intellect]. Thus, he does not posit another intellect to be the agent part of the [individual] soul, just as Plato does not, since he will have posited that the universals subsist *per se*. However, the intellect which has the nature of act according to Aristotle is the agent part of the human soul inclined for potential, like a light for colors. For it is immersed in matter and conjoined to dispositions in particular things, in order to abstract their species and deposit them in the impossible intellect, and projecting light on it, renders it actually intelligible.[47]

As Hasse notes, Pietro provides both Avicenna's and Aristotle's positions here. Avicenna posits that the agent intellect is not a part of the individual human soul, but is an external, independently subsisting intelligence that unites with an individual human soul when it has been perfected (here the third part of the potential intellect).[48] Aristotle thinks it is the power in each individual human that actively extracts universal knowledge from the material world and activates the potential intellect, comparing it to light.[49] Pietro himself seems noncommittal about this issue, alternatively calling this highest part of intellect *"in effectu"*, *"accomodatus"*, or *"acquisitus"*. His lack of commitment to a single position is apparent when he tallies the number of virtues; they total either 34 or 35, depending on whether one regards the agent intellect as belonging to each individual.[50] What is clear from this text is that Pietro did not hold the Averroist position that there is a single *potential* intellect shared by all individuals, as Nardi demonstrated. Even if he held the Avicennan position that all

**46** Hasse 2000: 174.

**47** *Conciliator*, diff. 57 (ff. 83v, col. 2, H–84r, col. 1, A): "Quando autem is actu intelligit intelligens se intelligere, intellectus est appellatus "in effectu" et tunc sibi coniungitur et unitur intellectus dictus "accomodatus" ab extrinseco. (Avicenna, 1 *De anima*). Vel acquisitus ut ab intelligentia, quam posuit "agentem" (*Meta.* 9). Et ideo intellectum non posuit alium agentem animae partem, sicut neque Plato, cum posuerit[is] per se universalia subsistere. Intellectus autem qui naturam habet actus secundum Aristotelem est agens animae pars existens humanae se habens ad potentialem, ut lumen ad colores. Ipsa nanque materiebus immersa et dispositionibus individualibus coniuncta, ut species earum extrahit et reponit impossibilem intellectum et super eundem proiiciens lumen, actu reddit intelligibile."

**48** See Hasse 2000: 177–179.

**49** Black 2010: 320–321.

**50** *Conciliatorr*, diff. 57 (f. 84r, col. 1, B): "Ex his itaque dilgenter computatis, apparet quod virium animae catena 34 aut 35 cum eisdem deservientibus constituitur annulis."

humans shared an *agent* intellect, this was neither a radical nor a heretical position.[51] We also learn that Pietro maintained that we have innate *per se nota* ideas, which distinguishes him from Aquinas, who denied such ideas.[52]

Regardless, it would be a mistake to take this text as a definitive statement about the intellect. Pietro's purpose in this part of the *differentia* is to define *virtus* by providing the alternative definitions characteristic of philosophy and medicine; he probably choses the reconciliation of Avicenna's three potential intellects with Aristotle as most representative of the position of a philosopher and because of its emphasis on development, while he does commit to a position on the agent intellect. None of this is controversial.

A second description of the intellect occurs in Pietro's commentary to *Problem* 14.1 in his *Expositio*. This problem asks:

> Why are those who live in conditions of excessive cold or heat bestial both in habits (*mores*) and in appearance (*visus*)? Are both due to the same cause? For the best complexion benefits the intellect, but excesses stupefy it. And just as they pervert the body, so they also affect the *complexion of the intellect*. (my itali).[53]

Many medieval scholars would have dismissed this text as a category mistake: an intellect cannot have a complexion. Nevertheless, even though Pietro is more than willing to disagree with the Aristotelian text elsewhere in his commentary, he does not do so here. Instead, Aristotle's parallel division of *mores* and *visus*, he says, should be interpreted as the acts of soul and appearance of the body, respectively.[54] That these two categories often share causes is the basis of physiognomy. This is especially clear in the case of animals, who teach us a lot about these kinds of links, adding that people who look like timid animals, such as deer and rabbits, have an excess of cold and are also timid. People with excessive heat are feral in *mores* and *visus*, having the habits and appearance of lions, wild boars, and the like.

However, the indication in the Aristotelian text of a link between *intellect* and complexion is a different issue, because animals do not have an intellect. To explain how we should understand the *complexionem intellectus*, he reviews Aristotle's definition of the intellect:

---

**51** Hasse 2010.

**52** Mahoney 1982: 608; Guldentops 2002.

**53** *Problemata*, 14.1: "Propter quid ferini secundum mores sunt et visus qui in superhabundantiis existunt aut frigoris aut caloris? Aut propter idem? Optima namque complexio et intellectivi confert, superhabundantie autem stupefaciunt. Et quedam corpus pervertunt, ita et complexionem intellectus."

**54** *Expositio*, 14.1: "Adhuc scias quod licet fere dicantur proprie animalia fortia sunt etiam agrestia, ut leo et pardus, unde descendit ferini. Nomen tamen hoc omne audiant brutum ut dicatur quod habentes excessum in frigore habent mores ferinos et visus: eo quod sunt timidi animalibus timentibus assiliati sicut cervi et lepores. Similiter habentes complexionem calore depauperatam superfluo sunt ferales moribus et visu. Sunt enim perfidi temerarii et similes apperentia talium animalium, ut leonibus et consimilibus. Et hoc est quod dicitur in Fisonomia: refertur ad leones, apros [sic] et huiusmodi."

> It must be known that since the intellect is four-fold according to the Philosopher, as appears in the *De Anima* 3, namely:
> 1) that of first principles, which come *per se* to the person having [an intellect] (*Metaphysica* 4);
> 2) the agent intellect, which makes everything;
> 3) the possible, or material intellect, which becomes everything;
> 4) the passive intellect, namely the corruptible imagination. Here [i.e. in this problem] the entire intellect can be understood, and especially the passible, in as much as its [i.e. the entire intellect's] activity depends on the passive intellect. For without it, there is no understanding.[55]

Notice that there is only one potential intellect in this list because Pietro is expositing Aristotle here, not Avicenna. He adds the sensitive power of the imagination (using "imagination" here to mean the whole complex of internal senses, rather than the individual faculty) and calls it the "passive (or corruptible) intellect", i.e. an intellect that can suffer from bodily affections. This is not particularly surprising, as it was not uncommon to refer to the imagination or the cogitation as an "intellect".[56] Occasionally, it could be used to explain what Aristotle meant by saying that animals have an intellect in *De motu animalium*.[57] For Aristotle's text here, this intellectual status for imagination was also useful for explaining *Problem* 14.1. Because the imagination is material, it explains why the text can refer to the "complexion of the intellect".

But Pietro gives the imagination greater emphasis than his contemporaries, arguing that we should not take Aristotle as identifying some detachable or distinct power, without which the intellect could still function, but that, as a practical matter, we should take the corruptible imagination for the "entire intellect", since the intellect's activity depends on it. Pietro explains that:

> And if there is no passible intellect, [the intellect] would have no organ in the body. Whence [I state] in the beginning of the *Physiognomy*: souls follow bodies and they, according to the authorities, are not unaffected by the motions of the body. Because of this, when there is a certain interior corruption, namely in the organ of imagination, the ability to understand and consider waste away.[58]

55 *Expositio* 14.1: "Notandum quod cum intellectus sit quadruplex apud Philosophum ut apparet tertio *De Anima*: puta (1) primorum principiorum que per se perveniunt ad habentem (4 *Meta*). (2) Intellectus agens, de quo dictum est quod est omnia facere ac (3) possibile ceu materialis quo est omnia fieri et quarto (4) intellectus passivus, ut imaginatio corruptibilis existens. Hic omnis potest audiri intellectus, et passibilis maxime, secundum quod quantum ad ipsius actum ex intellectu dependet passivo. Sine enim ipso non est intellegere."

56 Mahoney 1982: 607: e.g. Thomas Aquinas uses imagination in this way.

57 De Leemans 2006: 145–147.

58 *Expositio* 14.1: "Et si nullus intellectus passibilis, nullum habet organum in corpore. Unde in principio *Fisonome*: anime sequuntur corpora et ipse, secundum scripsas, non sunt impassibiles a motibus corporis. Propter quod, corrupto quodam interiori, ut organo ymaginationis, intelligere et considerare emarcescunt (primo *De anima* [Avicenna])."

Pietro's likely source for his claim that the intellect clearly suffers along with the body and therefore must have an organ is Galen's observation that people with head wounds lose various rational abilities.[59]

The final text that enumerates the parts of the intellect is in the exposition to *Problem* 4 in *Particula* 30, one of the two questions in the *Expositio* which Pietro judges to be about the intellect *per se*.[60] This text is particularly enlightening because Pietro does not follow the Aristotelian base-text closely; rather, by making several additions and qualifications, he directs the interpretation of the text to insist on his own priorities, namely the dependence of the intellect on physiology. *Problem* 30.4 asks:

> Why it is that we have a greater intellectual capacity when we are old, but that we learn faster when we are young? Is it because god gave us two instruments, by which we use external instruments: to the body, [he gave] hands; to the soul, [he gave] intellect? For the intellect is among those things which are in us naturally, as a tool.[61]

In the middle of the commentary, Pietro pauses to explain exactly what Aristotle means by the intellect as something that is in us "naturally, as a tool". To do this, he again specifies the different ways that intellect is meant in Aristotle's work:

> It must be known that 'intellect' is found to be taken in many ways in Aristotle:
>
> 1) the agent intellect (3 *De anima*), which makes [knowledge];
> 2) the possible or material, which becomes everything;
> 3) the imagination (also in 3 *De Anima*), [which is] the passive intellect and is corruptible;
> 4) the informative virtue of the embryo, which is called the intellect;[62]
> 5) Intellect is also the habit or faculty for recognizing principles, recognized as true without demonstration.[63]

---

**59** E.g. *Conciliator*, diff. 7 (f. 10r, col. 2,C).

**60** *Expositio*, 30.4: "Nota quod illud problema et 11 [30.11], qui sunt circa intellectum, et proprius secundum quod est circa sapientiam ad metaphisicam spectant."

**61** *Problemata* 30.4: "Propter quid senes facti magis de intellectu habemus; iuniores vero existentes citius adiscimus? Aut quia deus instrumenta in nobis ipsis dedit duo, in quo utimur exterioribus instrumentis: corpori quidem manum; anime autem intellectum. Est enim et intellectus eorum que naturaliter in nobis sunt, sicut organum existens."

**62** *Expositio*, 30.4: "Sciendum quod intellectus apud Aristotelem invenitur multipliciter accipi: aut per intellectu agente (iuxta illud Tertio *De anima*) quo est facere; vel possibili sive materiali, quo est omnia fieri; aut per imaginatione (de qua eodem) intellectus passivus corruptibilis est; aut per virtute informativa embrionis (de qua secundo *De Generatione Animalium*) talis autem est quod vocatur intellectus."

**63** *Expositio*, 30.4: Aut per quodam potentia nobis naturaliter insita qui et habitus principiorum, dicitur verorum semper existens, de quo *Metaphisicorum* Quarto scribitur. Quod omnem cognoscentem necesse sit ipsum habere, et venire ad habentem est necesse et non per demonstrationem acquiri. Primo *Posteriorum*. Talis enim notus omnibus natura non orbatis omnio ceu ianua in domo ut apparet Commen. *Metha*, 2°.

The last definition and the first three are familiar from *Problem* 14.1 above, but to them Pietro has added the fourth meaning in this list, that the intellect is the informative virtue of generation, which precedes the arrival of the soul (see Chapter 3), and which is "called the intellect" (*intellectus vocatus*) according to Pietro's interpretation of Aristotle's *De generatione animalium*.[64] This is a misunderstanding of William of Moerbeke's literal translation of a side comment that, in the process of the generation of animals, there is sometimes a divine element separable from matter and "what is called the intellect (*intellectus vocatus*) is like this". Pietro reads "*intellectus vocatus*", as naming a distinct kind of intellect and referring to the entire informative virtue. As Nardi argues, it was *this* unusual definition of intellect, which does seem to arise from material complexion, as "it often happens that the material lacks such an appropriate state and is unable to receive it",[65] which led to the accusation of materialism against Pietro.[66] He seems to understand that the informative virtue is not typically identified with "intellect", and he hastens to distinguish this intellect from the immaterial agent and potential intellects. It is in us "naturally", that is, from generation, "to distinguish it from the intellect later acquired by doctrine or study and to distinguish it from the separated intellect, which has no corporeal act".[67]

> By "natural intellect" I understand a habitual faculty of principles, naturally existing in us, so that from [that faculty] and by the abstraction of the agent intellect, the possible [intellect] is informed and perfected by intelligible species.[68]

It is through the later perfection of the potency of this intellect that we are able to "opine and know", and through which science and art and other rational virtues are attained.[69] It is the innate faculty of understanding principles and our natural tendency toward acquiring such knowledge.

64 Aristotle, *De Generatione animalium* 2.3 = 737a12–15 (Moerbeke transl., ed. Drossart Lulofs 1966: 54): "Geniture autem corpus in quo coegreditur quod animalis principii, hoc quidem separibile existens a corpore, in quibuscumque apprehenditur quod divinum (talis autem est [quod] vocatus intellectus), hoc autem inseparibile, hoc sperma geniture dissolvitur et spiritu repletur, naturam habens humidam et aquosam."

65 *Expositio*, 30.4: "Sed multotiens propter indispositionem non suscipiunt materie . . . . Sed susceptibilia aliquando sunt carentia dispositione suscipiendi ea."

66 Nardi 1912/1958: 4.

67 *Expositio*, 30.4: "Et exponit qualis sit intellectus huius dicens ipsum fore de numero eorum que naturaliter nobis inexistunt velut quoddam organum necessarium. Dicit autem que naturaliter nobis insunt ad differentiam intellectus acquisiti per doctrinam et studium vel ad differentiam intellectus separati, qui non est corporis."

68 *Expositio*, 30.4: "Per intellectum siquidem naturalem, audio huiusmodi potentiam habitualem principiorum nobis naturaliter inexistem secundum quod ex ipsa et intellectus agentis abstractione intellectus, possibilis deinceps informatur et perficitur intelligibilibus specierum."

69 *Expositio*, 30.4: "Dico autem intellectum quo anima opinatur et sapit. Primum tamen magis seriei coaptat subsequenti. Et quia dixerat intellectum hunc nobis naturaliter coniungi, assignant differentiam inter ipsum et scientias et artes, dicens scientiam ac artes quaecumque sint. Ille fiunt ex his que

Because the natural intellect is united to the body, Pietro makes the dependence of this intellect on complexion explicit with a noteworthy addition of his own. Where Aristotle states that the hands and intellect were given to us by God as natural organs for attaining the virtues, Pietro interjects that complexion should be added to these because of its importance: "With these two instruments [i.e. hands and the intellect], he [i.e. God] also gives a third, namely complexion, in virtue of which the foresaid two function."[70] If Pietro had simply wanted to equate the natural intellect in the text of the *Problemata* with a physical component of intellect, he could have easily identified it with "imagination" or "cogitation". But he is quite clear that he does not have the imagination in mind here. After all, the imagination would only develop later in life. Instead, he wants natural intellect to be something that is innate from the very beginning – the informative virtue – and which plants the seed for the later development of the other parts of the intellect that will come later.

The common thread in these texts is Pietro's insistence on a material condition for the functioning of the intellect, especially through the connection to complexion. The text of *Problem* 30.4, that "we have a greater intellectual capacity when we are old, but that we learn faster when we are young", could easily be interpreted to distance the optimal use of the intellect from the healthier physical qualities possessed by younger people. Instead, Pietro comments that by "old" (*senes*), Aristotle actually means "first old age" (*prima senectus*), a period Pietro places immediately after youth, when bodily passions begin to calm down, and "they understand more fully and perfectly" and not "second old age" (*secunda senectus*), when "the understanding and consideration deteriorate because of corruption in the interior [senses], namely the fantastical organ".[71] With this distinction, the time of greatest intellectual capacity is made to coincide with the healthy operation of the lower virtues, when the body has attained its most balanced complexion. Although Pietro is not more specific in this commentary, only saying that Aristotle distinguished two periods of life, *differentia* 19 of the *Conciliator* explains that the most temperate complexion is reached around age 35, which is presumably what he has in mind here – far from what most of us con-

---

in nobis existunt huismodi aut intellectus vigore. Quoniam rebus existentibus, scientie et artes non seipsas produxerunt inesse, immo ab intellectu causante."

70 *Expositio*, 30.4: "Cum his etiam duobus instrumentis dedit et tertium, complexionem scilicet, in cuius quidem virtute operantur pretacta 2° et precipue manus. Unde Avicenna Primo Prime [*Canonis*]: Donavit deus homini nobiliorem complexionem que in hic mundo foret possibilis invenire."

71 *Expositio*, 30.4: "Querit quare cum homines efficiunt senes – prima videlicit senectute – amplius et perfectius intelligunt. Non autem secunda ut senio eo quod intelligere ac considerare tandem emarcescunt quod quodam interius corrupto, puta organo fantastica (primo, *De Anima*)." And later: "Applicat secundo ad propositum, dicens quod similiter intellectus, qui est quedam potentia naturalis et in nobis non mox advenit a principio perfectus et actualis, sed tempore accidit precedente, puta in 'senectute prima,' cum vires corporee incipiant quiescere intellectualisque vigere. Non autem in senectute secunda in qua etiam intellectus debilitatur."

sider old.[72] It is at that age that our senses work best and that we are most prudent. This definition of *senectus* is an example of Pietro's consistent association of a temperate physical complexion with the operation of the rational virtues.[73]

One thing that is clear from these texts is that Pietro has little to say about the incorporeal agent and potential intellects. However, he always acknowledges that they constitute the highest part of the soul, almost as if he wants to forestall any charges of materialism. Indeed, he makes frequent comments to disavow any materialist reading of his work. Among many examples: in *differentia* 60 of the *Conciliator*, he describes the materialist theory of Alexander of Aphrodisias with disapproval, saying that Alexander holds that the possible intellect is educed from the potency of matter and thus is alone among Peripatetic philosophers in making the intellect corruptible.[74] Of course, this happens to be the exact charge against Pietro! In the *Physiognomy*, before getting into all the many ways the soul and body are connected, he remarks that "it is proven that [the agent intellect] is entirely separated".[75] Again, in the commentary to *Expositio* 30.4 discussed above, Pietro states that the intellect does not depend on the body for its "essence", before carrying on with the rest of his analysis.[76] Finally, with regard to the agent intellect in particular, he abruptly breaks off his fullest discussion of the topic about Averroes and the separated intellect as beyond the scope of the *Conciliator*.[77] He simply has little to say about it. In effect, all of these affirmations of an immaterial intellect, in its potential part and maybe an agent part, allow him to talk about the remainder of the "intellect" in the most material terms possible, and shield him, he hopes, from criticism.

---

72 Pietro seems to have created this age sub-division specifically to exposit this problem. At any rate, I have not found its use elsewhere. See: Paschetto 1984: 204–213, on Pietro's discussion of age, and Sears 1986 for medieval thought on periods of life.

73 See Chapter. 2: 57–60.

74 *Conciliator*, diff. 60 (f. 89v, col. 1, E): "Alexander peripateticus, Aristotelis prior expositor, nititur defendere, volens formas omnes ex elementorum mistione cum qualitatibus suscitari. Cui equidem positioni tamen innititur, quod et intellectum possibilem voluit de potentia materiae per actionem qualitatem educi. Et ideo solus inter caeteros peripateticos hunc posuit corrumpi."

75 *Phisonomie*, prologue: ". . . cum sic probetur omnino a corpore esse separata."

76 *Expositio*, 30.4: "Ipsa [anima intellectiva] enim non dependet a corpore saltem quantum ad esse . . ."

77 *Conciliator*, diff. 3 (f. 8r, col. 1, B–C).: "Quia igitur haec difficilior est tractatio quam negocium praesans exquirat, nunc tamen sufficiet dixisse."

# Chapter 5
# The Power of Things: Confidence and Incantations

According to Aristotle's *De motu animalium*, "imagination and thought have the power of physical things . . . and so we shudder and feel fear simply by conceiving an idea".[1] Aristotle's observation might seem obvious – if this were not true, the soul could not move the body. Pietro, however, uses it in support of a much broader variety of positions than Aristotle probably had in mind. In two questions among the practical problems in the second half of the *Conciliator*, he explains the way words, thoughts, and even external intelligences affect physical bodies, continuing his interest in the interaction of the material body with (almost) immaterial entities. The objects of Pietro's explanations range from the ordinary to the surprising, including the power of rhetoric, the internal psychology of the patient (what we might call a form of placebo effect), to action at a distance, celestial forces, and the intervention of angels and demons in nature, with little indication that he identifies a sharp distinction between these various phenomena.

*Differentia* 135 asks whether the idea of confidence in a physician can assist in actually healing a patient.[2] This question has a position of greater importance in the *Conciliator* than its placement as question 135 might suggest: it is one of four general questions introducing all regimens for curing illnesses, which occupy the remainder of the *Conciliator's* 210 questions, and he makes frequent reference to it elsewhere in his work. In *differentia* 156, a kind of companion question, he asks whether incantations can help cure a patient, and considers the effect of various kinds of magic (although this is not a word he uses) on the body.[3] In these questions, Pietro has no doubt about the efficacy of either confidence or spells; rather, the issue is exactly how they function.

These two questions have attracted more attention than most of his work.[4] Part of this attention is undoubtedly due to general fascination with all things magical, and Pietro wrote more openly and extensively about occult causality than any other

1 "Phantasia autem et intelligentia habent rerum virtutem. Aliquo enim modo species intellectur calidi et frigidi, aut delectabilis aut tristabilis, talis existit qualis quidem et rerum unaquequam, ut pretactum. Et ideo tremunt et timent, intelligentes solum." As quoted by Pietro in *Conciliator*, diff. 135, ed. Boughan 2006: 292. Aristotle, *De motu animalium* 7 = 701b17–23 (ed. Nussbaum 1978: 42). Latin ed. De Leemans 2011.

2 *Conciliator*, diff. 135 (ff. 191r–192v): "Utrum confidentia infirmi de medico conferat in ipsius salutem, necne." For the text of *differentia* 135, I use the edition in Boughan 2006: 2.282–304.

3 *Conciliator*, diff. 156 (ff. 212v–213v): "Utrum precantatio in cura conferat, necne." For the text of this question, I use the edition in Delaurenti 2012.

4 For example, Thorndike 1923: 902–904, Ottosson 1984: 264–268, Paschetto 1984: 235–247, Struever 1993, Boughan 2006: 1.66–110, Delaurenti 2007 and 2012, Giglione 2008, Compagni 2009, and Robert 2014.

https://doi.org/10.1515/9783110612639-006

scholastic author of this period.[5] Many doctors used charms of various types in their practice, but without necessarily interrogating their causality.[6] Beyond this, historians of medieval philosophy and science have been intrigued by what his explanations suggest about the limits of his naturalism.[7] Pietro tells us several times that his investigations are purely "natural", leaving out the miraculous and the divine, but his claims push the boundaries of naturalism beyond what most people today would be willing to accept.[8] Perhaps because of the obscurity of his prose in these questions, there has not been a consensus interpretation about his explanation of the phenomena. Some of his claims are clear enough, but others are so muddled by his undigested use of authorities and the disorder of his own writing that it is very difficult to be confident about his exact position.[9] For this reason, it will be useful to go through Pietro's positions in some detail, retaining some of the messiness that can be concealed in a neat summary of his text. It is also possible that this very obscurity has also contributed to the popularity of these texts, giving them an aura of mystery.

## Confidence as Cure

One of Petrarch's chief quarrels in his *Invectives against a Physician* (1355) is with "any doctor who pays more attention to empty eloquence than to his science".[10] Such physicians, he says, make the mistake of confusing rhetoric with the mechanical art of medicine. As sympathetic as we might be to Petrarch's complaint, Pietro and his contemporaries took for granted that it was medically beneficial for a patient to be confident in a doctor and that the doctor's words and demeanor were essential for the development of this confidence.[11] There was a range of opinions about why this worked: that patients who trust their physician are more likely to follow his instructions, but also the stronger claim that the positive emotions inspired by confidence actively aid in the healing process.

The subject of confidence was often addressed in commentaries on two of the most prominent books of the medical curriculum, the Hippocratic *Aphorisms* and *Prognostics*, both of which circulated in the *Articella* with commentaries by Galen.[12] Although not obviously about confidence, the second part of the famous first Hippo-

5 Urso of Salerno also wrote extensively about the causes of incantation in the late twelfth century. See Van der Lugt 2013.

6 See Kieckhefer 1989: 56–85, and Olsan 2003.

7 See Delaurenti 2007, and Robert 2014.

8 *Conciliator*, diff. 9 (f. 14v, col. 2, G–H). See Introduction: 8.

9 Boughan 2006: 1.74–75 also observes the messy nature of these questions, noting the "crude" use of quotations from authorities.

10 Petrarch, *Invective contra medicum* in *Invectives* (ed. and Engl. transl. Marsh 2003: 11).

11 See Salmón 2012 on the medieval tradition of this idea.

12 Salmón 2012: 196–201.

cratic aphorism was interpreted by Galen and later commentators as being about how to produce a trusting and obedient patient.[13] In the *Prognostics*, the topic is presumably knowledge of the future course of illnesses based on the interpretation of natural signs.[14] However, the Hippocratic text and Galen's commentary describe the value of future knowledge not so much in terms of its intrinsic utility, but as a rhetorical tool to impress the patient with the physician's knowledge and thereby build his trust.[15] Prognostication itself, even when it is certain, is not as useful for the attainment of health as the patient's belief in the expertise of the physician. The Hippocratic text reads:

> I hold that it is an excellent thing for a physician to practice forecasting. For if he discover and declare unaided by the side of his patients the present, the past, and the future, and fill in the gaps in the account given by the sick, he will be the more believed to understand the cases, so that men will confidently entrust themselves to him for treatment.[16]

Galen agrees that presenting a learned opinion about the future course of an illness is an ideal way to build the patient's confidence in the physician's ability.[17] However, the appearance of a well-informed opinion will function just as well.[18]

Among other leading authorities, Avicenna, in his *De anima*, takes the usefulness of confidence a step further and posits a more direct effect on the patient's health, claiming that confidence in the doctor is actually "more effective than what a physician does with his instruments and medicines", an opinion with which Pietro enthusiastically agrees.[19] By Pietro's generation in the Northern Italian universities, several physcians, along with the philosopher Gentile da Cingoli (fl. ca. 1300) turned to this issue with great interest and sought to explain the psychological benefit of confidence.[20]

---

**13** Hippocrates, *Aphorisms* 1 (ed. and transl. Jones 1931: 99): "Life is short, the Art long, opportunity fleeting, experiment treacherous, judgment difficult. The physician must be ready, not only to do his duty himself, but also to secure the co-operation of the patient, of the attendants and of externals."
**14** On medieval prognosis, see Demaitre 2003.
**15** See especially the commentary to the first piece of text by Galen and Taddeo (f. 195r–195v). For example: "It is from these [revealing signs of the past present and future] that the patient believes him to be a medicus and commits himself to him more certainly." Galen goes on to explain that creating confidence is more important than the correctness of the knowledge itself.
**16** Hippocrates, *Prognostic*, 1 (ed. and transl. Jones 1923: 7).
**17** Galen in Taddeo, *Expositiones*, f. 195r–v.
**18** Demaitre 2003: 776.
**19** Avicenna, *Liber de anima*, 4.4 (ed. Van Riet 1972: 2.64). "Attende dispositionem infirmi cum credit se convalescere, aut sani cum credit se aegrotare: multotiens enim contingit ex hoc ut cum corroboratur forma in anima eius, patiatur ex ea ipsius materia et proveniat ex hoc sanitas aut infirmitas, et est haec actio efficacior quam id quod agit medicus instrumentis suis et mediis." Pietro refers to this passage in *Conciliator*, diff. 135 (f. 192r, col. 1, B and f. 192r, col. 2, D).
**20** Siraisi 1981: 203–208. Gentile da Cingoli discussed a similar question about the power of intentional species on the body, asking: "Whether a sensible or intelligible species has the virtue of altering a body toward hot or cold." See Boughan 2006: 1.18–65.

In his own discussion of confidence, Pietro begins, as always, with the objections to their power. The primary difficulty is that nothing is able to act beyond its species (*nihil agit ultra suam speciem*). However, "health and confidence do not interact materially, since one is corporeal while the other pertains more to the soul and the intention".[21] "Health is a certain disposition caused by the temperance of hot, cold, humid, and dry, and by a balance of the quantity of the plasmas in number and position", and thus part of the natural virtue.[22] But confidence is a kind of "opinion" and belongs to the animal (or spiritual) virtue. All four of the objections deal with this theme in some form. Pietro also adds that sometimes the different levels of our being impede each other, so that "when confidence is strengthened as an animal virtue, natural virtue will be weakened".[23] This potential conflict of spiritual and corporeal health is reminiscent the strain of thought (discussed in Chapter 1: 28–29) which opposed medicine to spiritual health. However, Pietro's idea of the body-soul relationship is much more symbiotic than this objection suggests, and he never seriously considers the idea that confidence is not useful: the weight of "experience, authority, and reason" are too great to question its power. It is common, he says, to see a "sick person led to a dramatic recovery because of hope for a famous physician", noting that Galen says he was often successful while "doing nothing beyond others".[24]

Defining "confidence", Pietro says that it is "an intention strongly impressed on the occult apprehension".[25] Here, he draws upon the faculty psychology of the organic soul and the notion of "intentional existence" as they had developed in Paris in the second half of the thirteenth century.[26] "Intentions", "intentional species", or simply "species" are entities that are neither wholly material nor wholly immaterial. Their existence explains the diffusion of a network of forces and information, both visible and invisible, including the transmission of visual images from an object to an eye in sense perception, or other, hidden forces perceived by the internal (or occult)

21 *Conciliator*, diff. 135, ed. Boughan 2006: 282: "Sed sanitas et confidentia non conveniunt in materia, cum unum sit ens corporeum, reliquum autem animae et intentionis magis."

22 *Conciliator*, diff. 135, ed. Boughan 2006: 282–283: "Adhuc sanitas est quaedam dispositio causata ex contemperantia calidi, frigidi, humidi, et sicci, et coaequalitate plasmationis quantitatis numeri, et positionis (primo *Reg. San.*). Haec autem virtuti maxime inest naturali, ut quod eius monstratur operationibus; confidentia autem ad virtutem attinet animalem."

23 *Conciliator*, diff. 135, ed. Boughan 2006: 283: "Et confirmatur, quia cum virtutes se impediant, *De Anima* quarto Avicenne, fortificata confidentia virtute animali, debilitabitur virtus naturalis, ablato sibi calore instrumento nature, quo mediante suas peragebat actiones. Omnes enim virtutes tanquam proprio eo fulciuntur instrumento."

24 *Conciliator*, diff. 135, ed. Boughan 2006: 287–288: "[V]idemus non raro infirmos in grandem convalescentiam deduci propter spem medici famosi et diu affectati instrumentum quod magis aliis visum est non raro operari; medelae etiam operatione actuali extrinsecus, velut Galeni, de ipso testatur interdum, quod per me percepi non parum nihil interdum ultra alios operans."

25 *Conciliator*, diff. 135, ed. Boughan 2006: 285: "[C]onfidentia est intentio vehementer apprehensioni occulte impressa."

26 On the notion of "intentions" and "intentional existence", see Tachau 1999.

senses, as Pietro suggests can happen in the case of confidence. As Roger Bacon described them, these intentions are all around us and function as a kind of universal explanatory mechanism.[27] The power of intentions depends on the reality of their extra-mental existence, and although these intentions are almost entirely immaterial, they have enough force that, as they are received, they are able to physically "impress" the senses, however subtly, and transmit their information to an apt recipient.

The intention of confidence in the mind is formed from the apprehension of sensible species, which become gradually less material as they move from the external senses through the occult senses to the intellect, where an intellectual species is formed. An intentional species received by the particular senses is more material and impresses the patient more. This species "exists in a certain mixed manner", that is, between material and immaterial, "then [it becomes] more refined, in the common sense, and after that in the phantasy, the cogitation, and then the estimation".[28] The end result of this process of abstraction is a species in the intellect, which is "immaterial and universal, far removed from the particular thing on which it is to operate".

Emphasizing the physical nature of this process (at least until it reaches the intellect), the strength of the external intentions depends on physical characteristics of the patient, such as the age and disposition.[29] Someone "with a soft and delicate body" or who is very old or very young is more disposed to receive such intentions. The power of confidence within the patient who has received it also depends on the patient's complexion: "the confidence of those who have powerful imaginations, such as a burnt melancholic, impresses strongly, in contrast to that of a humid phlegmatic. For an act of an agent on a patient is impressed on one disposed to receive it."[30]

Alternatively, Pietro describes confidence as a passion of the soul. The passions (or emotions), are one of the six "non-natural" categories of things familiar from medical literature, and self-evidently cause material change, heating or cooling the body.[31] Aristotle is also clear in *De anima* 1 that the various passions, including joy, sadness,

---

27 Tachau 1997.

28 *Conciliator*, diff. 135, ed. Boughan 2006: 289–290: "Quod enim magis imprimit est quod sensu comprehenditur particulari. Hec enim species existit etiam modo quodam confuso, deiende absolutius, quod communi, postea quod fantasia, cogitativa, et deinceps extimativa, minime autem quod intellectu, eo quod intellectus et species eius sit immaterialis universalis, multum distans a re particulari signata, in qua est operari."

29 *Conciliator*, diff. 135, ed. Boughan 2006: 289: "Attendum tamen quod huius confidentia confert et penes corporis dispositionem variam et etatem, secundum quod aliquis mollis corporis delicati magis fuerit aptus in suscipere, [ac aetatis provectae vel contra, ac secundum quod intellectu], ac imaginatione, seu quacumque virtute interiori sensibili fuerit species confidentie recepta, comparatione illius quod actualius operatur exterius, cum actiofiat hinc in ratione satis differens." The bracketed line is mistakenly omitted in Boughan's edition. I have added it from the 1565 edition.

30 *Conciliator*, diff. 135, ed. Boughan 2006: 291: "Palin confidentia illorum qui fortis sunt imaginationis, ceu melancolici adustu, vehementer imprimit, quibus opponuntur humidi flegmati."

31 See Chapter 1: 20, for the non-naturals.

fear, and anger obviously affect the body, although Aristotle does not specifically include confidence here.[32] As Pietro explains:

> The body is moved by the passions of the soul (beginning of the *Physiognomy*) as is evident with loves, fears, sorrows, and the like [Aristotle, *De anima* 1]. All passions of the soul seem to be [also] with the body: anger, gentleness, fear, grief, confidence, as well as joy, love, and hate. For at the same time these [are felt], the body is affected . . . And confidence, as was said, is a passion of the soul, therefore the body is moved by it.[33]

However, confidence differs from anger, fear and the other passions of the soul, which cause more obvious and immediate change, because it is not in the sensitive part of the soul, but "is a certain passion (*affectio*) of the rational soul".[34] For this reason, he adds, confidence does not exist in animals.[35] Because of its rational nature, the force of confidence is less immediate and requires an explanation different from the other passions.

In fact, it seems that this confidence in the intellect cannot directly cause physical change, because "the [intentional] species of confidence in health, especially if it is in the intellect and very far from material health, does not bring health on its own".[36] Rather, once such an intention is formed, it must use intermediate means to cause real health, since intellectual species are only capable of interacting with other abstractions: the intellect can only contemplate healing the abstract "man", not the individual Socrates or Callias (a favorite example of Pietro's).[37] However, we know that there must be a way for the thought of confidence to cause a cure, and Pietro invokes Aristotle's well-worn remark from *De causa motus animalium* that, "phantasy and intelligence have the power of things. For in some mode the species in the intellect of hot or cold or delight or sorrow, exist like [physical] things."[38]

---

**32** Aristotle, *De anima* 1 = 403a4–27 (ed. and transl. Hett 1936: 13–17).

**33** *Conciliator*, diff. 135, ed. Boughan 2006: 294–295: "Rursus 'corpus compatitur a passionibus anime,' in *Phisionomie* initio, velut apparet circa amores, timores, tristitias et huiusmodi. Uunde prohemio *De anima* [Aristotelis]: 'Videntur omnes anime passiones fore cum corpore: ira, mansuetudo, timor, misercordia, confidentia, adhuc gaudium, et amare et odire. Simul enim hiis cum anima aliquid compatitur corpus.' Quod apparet: 'Nullo enim terribili imminente extra in passionibus efficimur quandoque timentes,' sed confidentia, ut dictum, anime extat passio; ergo patietur ab ea. . . ."

**34** *Conciliator*, diff. 135, ed. Boughan 2006: 296: "Rursus confidentia est anime sit rationalis quedam affectio. . . ."

**35** *Conciliator*, diff. 135, ed. Boughan 2006: 296: "Omnem namque opinionem sequitur fides, nulli autem inest bestiarum."

**36** *Conciliator*, diff. 135, ed. Boughan 2006: 290: "Species tamen confidentie sanitatis, et si valde distat a sanato, per se quidem non conferens ei, tamen per accidens et modo quodam applicatur intermedio."

**37** *Conciliator*, diff. 135, ed. Boughan 2006: 290.: "Actus etenim et operationes circa singularia existunt; *Metaphysice* initio. Non enim hominem medemur, verum Sortem aut Calliam." See Chapter 1: 32.

**38** *Conciliatorm*, diff. 135, ed. Boughan 2006: 292. Aristotle, *De motu animalium* 7 = 701b17–23, as cited in the first note of this chapter.

To accomplish this return from the abstract state in the intellect to the sensitive part of the soul, the intention of confidence reverses the path by which it was formed and instructs the soul's lower levels. Apparently because the intellectual species was originally derived from more material species (phantasms in the imagination), it retains some connection to its origin:

> For when the intellect understands, it is necessary that a phantasm is seen. A phantasm is a likeness of a corporeal thing. And thus, confidence existing in the intellect in some way will have agency, since it is derived from the altering of the sense. For nothing is in the intellect which was not first in the sense (Aristotle, *De Anima* 2).[39]

Next, it can go back through the levels. Once it gets to the imagination (which deals with particular images), the concept has returned to the realm of particularity and materiality where all real healing occurs.

> It should be noted that confidence in health is led by that path from particular sensations of health to the fantastical virtues or to the intellect, and by returning from that [path], it aids in [real] health. Since it was in the intellect, by reversing, it is led back to the powers of the fantasy, as [first] to the estimative, then to the cogitative and imagination. And then, as it is made material and particular, it is restored even more as an instrument of nature, like heat, and then it excites itself, especially that seated in the heart, by whose accident, that is the vital spirit, it is taken for altering, loosening, and destroying what there was of the illness.[40]

Finally, Pietro argues that it is actually more effective than material cures, because it acts more gently:

> Furthermore, whatever changes less by sudden alteration, is able to greatly aid health, in so far as nature does not withstand sudden change. But confidence in health accomplishes this because it subtly induces alterations, for it does not operate violently as do many material [treatments], applied internally or externally.[41]

---

**39** *Conciliator*, diff. 135, ed. Boughan 2006: 290–291: "Cum enim intelligit intellectus, necesse est aliquod fantasma speculari. Fantasma quoque est similitudo rei corporalis. Et ideo confidentia existens in intellectu modo aliquo habebit agere, cum dependeat a sensu magis alterante. Nihil enim fit in intellectu quin prius fuerit in sensu; secundo *De anima*."

**40** *Conciliator*, diff. 135, ed. Boughan 2006: 301–302: "Iuxta hunc quoque modum ponendum est quod confidentia sanitatis, via qua ex sensationibus particularibus sanitatis deducta est, in virtutes fantasticas aut in intellectum revertendo ab illo, conferat ad salutem eandem. Cum enim fuerit in intellectu, revertendo reducitur ad vires fantasticas, ut extimativam, deinde ad cogitativam et imaginationem. Et tunc, velut materialis et particularis, instrumentoue nature, ut calori, amplius reddita, simul excitat ipsum, maxime prout fundatur in corde. Quo accidente, id est spiritus vitalis suscitatur ad alterandum, resolvendum, et destruendum, que fuerant egritudinis. . . ."

**41** *Conciliator*, diff. 135, ed. Boughan 2006: 295: "Amplius quodcumque minus subita alteratione permutat potest conferre ad salutem maxime, eo quod natura non substinet repentinas mutationes. Sed hoc operatur sanitatis confidentia, cum suaviter inducat alterationes. Non enim sic violenter operatur, ut materialium plura intus vel extra applicata. . . ."

It does this by directing spirits in the healing process, in that there is a great range of potential in the illnesses it can cure. "Either the illness will be cold, and the heat will cure it by contrariety, or the illness will be hot and the heat will destroy it by forcing it to evaporate."[42]

When we think back to the initial objections to the idea that confidence helps in a cure, we see that Pietro has addressed them in the following ways: first, he clearly maintains that the different virtues in the soul have a symbiotic relationship, i.e. we do not need to worry that a healthy intellect will impede our other virtues. Secondly, he insists that the power of confidence is not a case of something completely immaterial causing change in matter. Although confidence as it exists in the rational soul in an "intentional" and almost entirely abstract way cannot cause change in gross matter on its own, it does work its way back through the levels of the soul, from highest to lowest. The intentions in the intellect are almost, but not quite, immaterial. They affect the imagination in the sensitive soul; the phantasms in the imagination are in turn more material and can then effect more material change.

## Incantations and the Power of Words

Enchantments did not have the same universal support from authorities as the notion of confidence, but they still had a long history in medical practice. When physicians attempted to explain the cause of the effectiveness of spells, the connection between the spell and the psychology of the patient, particularly the patient's trust that the spell will work, is immediately apparent. One of the most influential texts in this regard is Qustā ibn Lūqā's (ca. 820–ca. 912), *De physicis ligaturis* (On Physical Ligatures), which describes a variety of objects which are hung or tied to the body to produce a cure.[43] With these, ibn Lūqā says repeatedly, the patient's attitude and psychology are of greater value than the internal properties of the ligature. In the late twelfth century, the physician Urso of Salerno produced the first work in the Latin West which discussed the natural causes of incantations. He explains their effectiveness in terms not unlike what we find in Pietro d'Abano: the soul of the doctor (in this case the "aura" transmitted in the doctor's spirits) and their interaction with the confidence in the patient's soul produce health.[44] In the second half of the thirteenth century, the source of occult properties in various objects was a common matter of discussion

42 *Conciliator*, diff. 135, ed. Boughan 2006: 293: "Calor enim sicut instrumentum nature curat egritudinem. Nam aut egritudo erit frigida, et in contrarietate quam habet ad ipsum curabit; aut calida, et tunc movebit spiritum ad ipsam destruendum, evaporare faciendo eius materiam, et resovendo; *prima secunda*. Haec autem agit confidentia, ut ostendetur."

43 On this text, see Wilcox and Riddle 1995, which includes a Latin edition of ibn Luqa's text: 26–48, and Boughan 2006.

44 On Urso's text, see Van der Lugt 2013, which includes an edition and English translation 334–346.

among natural philosophers, including Albertus Magnus, Thomas Aquinas, Roger Bacon, and others.[45] However, many learned physicians remained skeptical of the value of incantations, which lack the physical presence of objects.[46] As Pietro d'Abano's contemporary Arnau of Vilanova (d. 1311) understood it, most occult properties cannot be discovered by reason and incantations in particular, are "clearly beyond plausibility".[47]

When Pietro turns to the subject of enchantments in *differentia* 156, one type of objection to their ability to affect the body is similar to the objections to the power of confidence: spells, and words in general, belong to a different realm of existence than physical health. Anything that is able to heal, must be able to interact in matter, but an enchantment and health do not do this, and nothing acts beyond its species.[48] We have seen that this objection is not difficult to overcome. However, in the case of enchantments, the medical authorities are much more negative than they were for confidence. Galen, Pietro recognizes, does not count spells among the categories of medicine;[49] further, he also "derides physicians who incant and conjure with herbs".[50] However, like confidence, Pietro regards the evidence for their effectiveness to be incontrovertible.

In response to the objections, first he offers a surprising example from the beginning of the *Illiad*, where he recounts that the Trojan priest Chryseus (whose abducted daughter he identifies as "Brisies" not "Crisies") successfully used an incantation (along with reminders of how many sacrifices he had made) to petition the sun to cause a plague among the Greeks with his arrows (or rays/*radio*).[51] Offering more detail than one might expect, he explains that the sickness first affected the mules, then the white dogs, then the men, before the prophet Calchas identified the cause. Pietro seems to interpret this as an example of astral magic. He is back on firmer ground in

---

**45** Weill-Parot 2010.

**46** McVaugh 2003.

**47** McVaugh 2003: 336.

**48** *Conciliator*, diff. 156, ed. Delaurenti 2012: 68: "Rursus quod sanat, habet alterare, cum sanitas reducatur eo. Sed incantatio non alterat, quia quae huiusmodi sunt, in materia communicare oportet, quod non incantationi adest et sanitati, ut praetactum. Non igitur sanationi praecantatio confert."

**49** *Conciliator*, diff. 156, ed. Delaurenti 2012: 68: "Adhuc *primo quarta*: Res medicationis rebus perficitur tribus, puta nutrientibus, potione seu medicina et manuali operatione. Unde Galenus *Paterniano*: 'Totius corporis medicina in hec tria redigitur: medicamentum, ferrum et ignem.' Sed incantatio nulla est istorum, ut monstratur ex prosecutione ipsorum."

**50** *Conciliator*, diff. 156, ed. Delaurenti 2012: 67: "Galenus VI *Simplicium singularium*, deridet medicos incantates quosdam et coniurantes herbas."

**51** *Conciliator*, diff. 156, ed. Delaurenti 2012: "Chrisses sacerdos, propter Briseidam filiam suam raptam per Agamemnonem rediens secus litus maris sibilantis, ea quidam non obtenta, rogavit incantatione Solem, dicendo qualiter sacrificavit sibi et sacrificabit, si de Grecis faceret ultionem, ita ut sol concitatus immittens arcu sagittas eius seu radio, causavit pestem in mulos primitus et canes albos, et demum in homines, ut cogretatur Agamemnon Briseidem reddere sacerdoti, Calcade siquidem Testoride pestis huius causam assignante." See *Iliad* 1.38–118.

his final two responses to the objections, when he refers to the ubiquitous quote from Aristotle's *De motu animalium* about the action of thoughts on the body and to the role of confidence in healing, explained in *differentia* 135.

Defining "incantations", he says that "Socrates said that incantations are words that deceive human souls."[52] And further, "[p]recantation is a wonderful speech (*oratio*) delivered with great passion (*affectione*) to aid the person being enchanted, especially when they are confident".[53] It is actually more effective, he says, if the words are largely meaningless (*non significativorum*) and about "wonderful and hidden things" (*de miris et occultis*). All of this will make the patient "judge that the enchanter can better help".[54] As we will see, the meaning of "precantation" is very broad: it ranges from spoken formulae to liturgical rites, to appeals to the planets, to words written on objects, and more. Further in the text, Pietro offers a rather disorganized list of examples of the effects of incantations as empirical evidence that they work.[55] These include the sacrament of the Eucharist; the *ars notoria*;[56] killing a bull by speaking words in its ear (as an enchanter did in front of St. Sylvester, who then brought it back to life);[57] the movement of consecrated bread or a psalter or sieve for discover-

52 *Conciliator*, diff. 156, ed. Delaurenti 2012: 70: "Propter primum quidem sciendum, quod scribitur Socratem dixisse incantationes fore verba animas decipientia humanas." Pietro takes this definition from ibn Lūqā (ed. Wilcox and Riddle 1995: 33).

53 *Conciliator*, diff. 156, ed. Delaurenti 2012: 70: "Ut autem in praeiacenti sit loqui materia, praecantatio est oratio admiranda affectione in subsidium incantati, praecipue confidentis explicata."

54 *Conciliator*, diff. 156, ed. Delaurenti 2012: 71: "Dicitur autem oratio ad discretionem terminorum non significativorum, que, cum fuerit de miris et occultis, amplius eam reputans incantatus, eidem conferet magis. Et ideo quantum possibile occultatione utendum: 'deitatem enim minuit, qui secreta mistica vulgat.' Ars notaria praecantans etiam cum grandiori diligentia debet ipsam promere, ut tam ab ipso [i.e. incantando], quam ab incantato dignissima reputata possit amplius conferre. Cum enim fuerit de ipsa confidens amplius, ut ita eundem iuvabit maxime, iuxta illud differentie 135: "ille plures sanat egritudines, de quo plures confidunt."

55 *Conciliator*, diff. 156, ed. Delaurenti 2012: 77–78: "Propter secundum huius sciendum quod experientia potest monstrari, et demum ratione persuaderi precantationem conferre. De hoc enim in simile et proposito empirie sunt quam multae, ut aperte illud summum sacramentum cum aliis multis ostendit eucharistiae. Nomina etiam id confirmant divina notorie artis et eutuntice. Magis etiam verbis prolatis in auriculam tauri, cum incantator prosternerit in mortem in praesentia Silvestri [nostri in 1565 ed.!], hunc et reviviscere denuo effecit. Hoc etiam confirmat motus panis ad furtum inveniendum aut psalterii et cribri, verba etiam quedam in aurem cuiuscunque hominis prolata: satagit eum tibi, quecunque petieris elargiri. Incantatione similiter pronunciata vel scripta ne unquam pandant qui cruciantur, ea etiam super gladium ambulant acutissimum et prunas vivaces, digitorumque uno applicantes alterutrumque hominem aut pondus sublevant in altum gravissimum. Serpentenque stupefaciunt ut non ledat. Equos furibundos carmine quietant eosque castrant cum aliis, {canes mulcescunt latrare cessantes}, immobiles eadem redditi. Hoc etiam modo trucidati evadunt periculum."

56 For the *ars notoria*, a book of ritual magic originating in Italy in the late-twelfth century, see Page 2013: 112–129.

57 The 1565 edition of the *Conciliator* replaces the word "Silvestri" with "nostri", which causes Thorndike 1923: 903, and Ottosson 1984: 265 to report that Pietro claims to have witnessed (*in praesentia nostri*)

ing a thief; and words that make someone want to give you whatever you ask for. Similarly, spoken or written incantations allow people to withstand torture, to walk on a sharp sword edge or hot coals, or to lift a man or heavy weight to a great height with one finger. Incantations stun snakes so they do not attack, calm wild horses to be castrated, and stop dogs from barking, rendering them motionless and enabling people to escape. They make someone unable to sleep, continually imagining the enchanter, and they cause nocturnal pollution. An arrow stuck in a bone can be easily pulled out without pain, cancer is destroyed, worms are killed, toothaches are cured. With certain characters, a flow of blood can be stopped. Epilepsy can be calmed by speaking the words of the names of the magi in someone's ear or with a more elaborate ritual he attributes to Constantine (actually ibn Lūqā). Kidney pain can be cured with characters and song.[58] In the middle of this list of very specific examples of incantation, he adds that the planets can also be implored for help following the correct procedure.[59] The full list sounds like the result of a lengthy brainstorming session.

Considering the sharp difference of these examples and the lack of specific explanation for each of them, it is difficult to say if he really believes every item here. Certainly, he believes many of them and regards them collectively as evidence for the obvious efficacy of spells. However, Pietro also wants to situate himself between materialist skeptics, who do not believe in enchantment at all, and credulous people who will believe anything. He speaks of all these people derisively:

---

a man being brought back to life by an enchanter in his presence! Obviously, this changes the force of this observation significantly.

**58** *Conciliator*, diff. 156, ed. Delaurenti 2012: 80–81: "Incantatione quoque, quis dormire non valet, representationem continue incantantis imaginans, ipsa etiam polutionem inducit in somniis. Quod etiam futurum de aliquo taliter apperit. Sagitta etiam ossi adeo infixa ut alio nequit ingenio extrahi, digitis lateraliter duobus facile ac sine dolore dulciter applicatis foras educitur. Cancerque mortificatur, lumbrici occiduntur {dolorque dentis curatur}. Sanguis ea et charactere undecunque fluens sistitur. Paroxismus epilepsie quietatur nominibus Magorum in aure prolatis, aut portatis super se cessare cogit omnino. Epilepticus sanatur et lunaticus, Constantinus cum aliis, 'cum pater aut mater in Quatuor temporum feria quarta eum ad ecclesiam deducat, et feria sexta, et deinde sabbato super ipsum cantetur et suspendatur illud dominicum evangelium: hoc genus demoniorum non eicitur nisi orationibus et ieiuniis ipsam.' Etiam dolor dentis mox tollitur. Charactere similiter et carmine passio curatur renalis. Futurum prenoscitur et quod abest, ut edocet geomantia, taliumque multa."

**59** *Conciliator*, diff. 156, ed. Delaurenti 2012: 79–80: "Prima similiter astrorum oratione placantur et in subsidium concitantur nostrum ut orationum epilogus insinuat planetarum. Unde *Albumasar in Sadan*: 'Reges Grecorum, cum volebant obsecrare deum propter aliquod negocium, ponebant Caput Draconis in medio celi cum Iove aut in aspectu ab eo figura amicabili, et Lunam coniunctam Iovi aut recedentem ab ipso et coniunctionem cum domino ascendentis petentem, aut domini ascendentis coniunctionem cum Iove petentem, adhuc autem cum Capite amicabili figura. Tuncque dicebant ipsorum petitionem exaudiri.'"

> [S]ome say incantations do not work, either because they are gross skeptics, not thinking any acts happen except through material and very gross contact through the mediation of very obvious primary qualities. This happens to many of them because of their great immersion in corporeal matter, so that they are not able to conceive of the present, however obvious, and even less the future.[60]

These people are not just wrong; they are morally repugnant, mixed up in shameful material things and not taking sufficient pleasure in their rational souls. As for the people who will believe anything, he quotes a passage from Galen, who describes physicians who "wrote superstitious words and stories, which little girls (*mulierculae*) and old women (*vetulae*) are wont to tell, in the manner foolish Egyptians".[61]

Enchantments, he says, work either by internal or external causes. The internal causes are twofold. One cause is the soul of "a particularly passionate enchanter, whose soul is so elevated that it is able to dominate mundane matter and transform it to his will by the energy received from the intelligence".[62] Alternatively, the cause is the soul of the person being enchanted, "insofar as he is confident, so that he is helped not a little by the strength of his estimative faculty".[63] For further explanation of these internal psychological causes, he refers the reader back to *differentia* 135 on confidence, where he judges he has explained this sufficiently.

---

**60** *Conciliator*, diff. 156, ed. Delaurenti 2012: 73–74: "Propter secundum quidem sciendum quod aliqui dixerunt precantationem non conferre, aut quia grossi increduli non putantes aliquas actiones provenire nisi materiali et grossiori quodam contactu mediantibus qualitatibus primis sensibilioribus. Valde licet omnis actio inde tandem apud Aristotelem consurgat, differentia 60. Quod accidit propter immersionem plurimam ipsorum [in] materiae corporali, unde nec praesentia utcumque denudata, futuraque minus possunt concipere. Quare Avicenna, Metaphisice nono 'Quia nos in hoc nostro secola et corpore demersi sumus in multa turpia, non sentimus delectationem animae rationalis cum aliquid apud nos fuerit de causis eius.'"

**61** *Conciliator*, diff. 156, ed. Delaurenti 2012: 74–75: "Alii autem eas spernunt {nacti in phylosophia communiori omnium causas fere querentes mathematicas. Nonnulli vero} rationaliter {has contempnunt}, eo quod in eorum non cadebant artem sensibilem et entem et apertam, in qua actus considerantur manifestiores, qualitates parum excedentes primarias, et maxime cum incantent anima carentia rationali vel sensitiva. Ceu Galenus in sexto prealegato, detestans Chamachirum et Bamachirum, cunctis Dioscoridem praeferens, eo quod scripsissent superstitiosa verba et fabulas quas narrare consueverunt mulierculae ac vetulae ad modum stultorum Aegyptiorum fascinationes dicentium. Coniuraverunt enim medicinas, asperserunt et suffumigarunt divinitus, cum eas [i.e. herbas] evellerent a terra ut ipsas collo vel alibi, iuvamenti causa in modum suspenderunt filateriarum. Quod omne falsum est, ac stultum, et medicinae artis detestativum. Hii enim inhaerentes huiusmodi qui artis dereliquerunt propria."

**62** *Conciliator*, diff. 156, ed. Delaurenti 2012: 82: "Aut ex parte incantantis affectuosi, cum eius fuerit anima in tantum elata ut materie possit huic dominari mundane, ipsamque vigore intelligentie suscepto ad nutum transmutatre."

**63** *Conciliator*, diff. 156, ed. Delaurenti 2012: 82: "Vel ex parte incantati, quamplurimum confisi, ut ex vigore siquis iuvatur estimationis non parum."

For external causes, he divides them into a universal cause, which is God, or particular causes, which are further divided into whether they are joined to matter or not.

> The more universal [cause] is the first of causes, by virtue of which an *oratio* or prayer takes effect, since it contains your concerns. The first cause, through his sheer kindness, since he is a supernatural agent, is able to operate in the inferior world, through his will, without motion or transmutation.[64]

God is able to work in this way, he says, "whatever a Peripatetic says", indicating that although he realizes such direct action disagrees with Aristotle, he is not going to let that bother him.

Particular external causes, independent of matter, are either angels or demons. If it is benign, it is an angel, "who is appointed as our administrator and brings our prayers to God". The Testaments are full of examples of this.[65] The demonic causes are more problematic:

> If it is depraved, it is a demon, or spirit, whose effect is tremendous, since there is no power on earth to compare to it. It often obeys incantations and sacrifices, as it strongly desires to deceive, so that it appears to bring about many things, and it is especially obedient to silly women and the simple, since it is able to deceive them more easily than a wise man.[66]

Pietro has witnessed this sort of thing, he adds, saying, "I have seen a pact or effect follow from the incantation or conjuration of [unlearned people] quite often, which does not happen to someone wiser."[67] Next, when the cause comes from a moveable, corporeal body, the bodies may be either incorruptible or corruptible. If it is moveable and incorruptible:

> this is an intelligence or soul of the heavens. Their successive motion aids the incantation, since they attend to our concerns in a certain manner (as has been mentioned). So, this intelligence helps or guides by means of its planet or star in various ways depending on what was chanted.

---

**64** *Conciliator*, diff. 156, ed. Delaurenti 2012: 82–83: "Universalior quidem ceu causarum prima, cuius quidem virtute oratio seu precantatio suscipit effectum, cum ipsa quodammodo nostram sollicitudinem habeat, ut ostensum differentia 101. Potest namque Primus sua mera benignitate, cum sit agens supernaturale, per voluntatem absque motu et transmutatione in hec inferiore operari, quicquid dicat Peripateticus differentia submissa, cum ipse sit summum bonum de cuius ratione fit 'sui fore diffusivum ac communicatum' aliter enim bonum minime foret summe."

**65** *Conciliator*, diff. 156, ed. Delaurenti 2012: 84: "Quod si primum, aut est benigna, ceu angelus qui noster ponitur administrator a divinis et nuntius orationes deferens ante Primum, sicut etiam sensit Avicenna, decimo *Metaphisice*. De quo sunt non parum Testamenta repleta."

**66** *Conciliator*, diff. 156, ed. Delaurenti 2012: 84–85: "Si vero depravata, ita demon vel spiritus, cuius effacacia permaxima cum 'non sit in terra potestas que sibi valeat comparari,' hic quidem multum incantationibus et sacrificiis obedit ut decipiat concitatus vehementer, aut apparaet multa perficiens, et precipue mulierculis obediens simplicioribus, [quia] eas enim celerius sapiente fallere valet."

**67** *Conciliator*, diff. 156, ed. Delaurenti 2012: 85: "Quare has video votum et effectum consequi ex earum incantatio seu conjuratione utplurimum, que sapientiori non accidit, quantumcunque illis operetur perfectius."

> Quite a few magi observe this, invoking Jupiter, Saturn, or another planet, by the name of the intelligences ruling it, when they were fortunately or otherwise well-disposed.[68]

Returning to his earlier example from the *Iliad,* he adds that "Chryses the priest certainly thought this". Again, he adds that he has personally witnessed (*ut expertus sum*) people being made to love or hate with a chant invoking these planets.

The cause united to a "corruptible, moveable, body is the intelligence called agent by Avicenna", possessed by people with particularly powerful souls, able to act on external matter, and he refers the reader back to *differentia* 135 again, along with 101 (on this see below).[69] Finally, the causes might not really be causes at all, but just coincidences, in which case the incantation does not really do anything, but merely seems to.

Whatever the power of any particular cause, Pietro always emphasizes the need for the enchanter and the recipient of the spell to be properly disposed. A skeptical patient will not be persuaded or enchanted, regardless of the rhetoric or the spell. An incantation works best on someone who is "desirous, as hopeful as possible, and predisposed in every way that the action [of the spell] occur in the prepared matter".[70] The enchanter should be "astute, believable, passionate, and of a strong mind".[71] "And this makes sense, since an incantation is something intentional, it does not work unless the aforesaid things occur, since the action of any agent is on a patient predisposed to accept them."[72] He adds that spells are more effective if the patient is asleep (but presumably prepared for the procedure), because it can directly alter the more mutable animal virtues when the rest of the body has ceased motion. In this case the intentional species directly impress the internal senses.

Several issues in his account of these phenomena have attracted particular interest from modern historians, all of them related to the limits of Pietro's naturalism.

---

68 *Conciliator,* diff. 156, ed. Delaurenti 2012: 86–87: "Si vero applicatur huius substantia mobili aut incorruptibili, ceu intelligentia et anima celi, vel altera tamen apud Alexandrum et Themistium, et hoc dupliciter. Aut enim confert incantatione alteratur mota, cum ordinetur ad nostrum sollicitudinem modo ut tactum quodam sitque intelligens, ita ut eius orbe ac astro conferat aut officiat secundum quod diversimode fuerit incantata. Quod observant magorum non pauci invocantes Iovem, Saturnum, vel alium nomine intelligentie ducentis eundem, differentia 9, cum feliciter fuerint vel aliter dispositi. Quod equidem Chrisses Sacerdos sensit. Et, ut expertus sum, ipsos in amorem proprie ac odium non leves continuato adducunt carnine."

69 *Conciliator,* diff. 156, ed. Delaurenti 2012: 88: "Si autem substantia mobili adhereat corruptibili, ceu Intelligentia dicta agens ab Avicenna, quid de hoc in se ac in propsito sentiendum differentia expositum 101 et 135."

70 *Conciliator,* diff. 156, ed. Delaurenti 2012: 72: "Incantandus vero, avidus, sperans quammaxime, ac dispositus omnimode ut actio concidat in materiam preparatum."

71 *Conciliator,* diff. 156, ed. Delaurenti 2012: 72: "Amplius precantator debet esse astutus, credulosus, affectuosus, anime fortis impressive. Et ideo pertransmutat et alterat quod est maxime permutabile, ceu virtutem animalem, et maxime sompniis, cum motus reliqui tunc cessant corporei."

72 *Conciliator,* diff. 156, ed. Delaurenti 2012: 72: "Et merito quia, cum incantatio sit quid tamquam intentionale, non agit effacaciter nisi intervenient predicta, cum 'actus agentium sit in passum et susceptivum predispositum.' Et ideo [precantatio] pertransmutat et alterat quod est maxime permutabile ceu virtutem animalem, et maxime somniis, cum motus reliqui tunc cessent corporei."

One issue is the power contained in the words themselves. Is it ever the power of the spoken or written words alone that cause the enchantment to work? Pietro raises this possibility, but quickly dismisses it, saying "an enchantment does not cause a cure on its own (*inquantum ipsa*), but only in so far as it receives power from the person uttering or conjuring or from [whatever] is establishing or causing it".[73] Otherwise anyone could cast spells effectively, which is not the case. When unqualified people seem to successfully cast spells, this is the power of demons.

This causal role of demons in nature is a second issue and a more fraught one. Some historians have sought to downplay the role of demons in Pietro's causality, drawing a clear distinction between natural and demonic causation. Some note that there is a general tone of skepticism throughout *differentia* 156, or that Pietro's thoughts on the matter are muddled enough that it is impossible to say how great a role he grants to the direct intervention of demons in nature.[74] Others have identified a more coherent rational philosophical position in Pietro's texts, which seeks to systematically reduce the role of demons in favor of natural causation. Delaurenti, in particular, argues that "Pietro d'Abano refuse d'établir un lien causal entre les démons et les incantations", in his naturalistic explanation.[75]

However, even if demons only fool *mulierculae* into thinking they are casting effective spells, this does not change the fact that the demons are still the real causal agents.[76] Nobody at the time would have denied the existence of demons, nor their intervention in the natural world, however troubling that might be to modern notions of naturalism. More to the point, there is no reason to dismiss what Pietro tells us with his own words: that demons are one of the causes of the enchantments he has just itemized. Instead of trying to expel all demonic causation from spells, it is more plausible that he is defending the use of magic for learned physicians like himself, while accusing everybody else of consorting with demons.

A final question has been Pietro's attitude toward the possibility, associated with Avicenna's psychology, that some particularly powerful souls might be able to act directly on external bodies without physical contact, contradicting Aristotelian physics.[77] Towards the end of his discussion of the formation of confidence in *differentia* 135, Pietro reports Avicenna's claims about these souls:

---

73 *Conciliator*, diff. 156, ed. Delaurenti 2012: 76: "Propter tertium quidem sciendum primitus quod non confert praecantatio in cura, inquantum ipsa, sed demum inquantum talis prout vim scilicet, recipit a proferente, vel coniurante aut ab instituente seu causante."

74 Thorndike 1923: 903, remarks that "All of this sounds rather skeptical on Peter's part." Ottosson 1984: 265–267, thinks this is pretty muddled and definitely inconsistent with his Aristotelianism elsewhere.

75 Delaurenti 2012: 45.

76 As Giglioni 2008, points out, the desire to separate demonic from natural causality is primarily a modern concern, and not Pietro's.

77 On Aristotle's principle that change requires physical contact, see e.g. *Physica* 3 = 202a2–12 (ed. Wicksteed and Cornford 1957: 205–207).

> Avicenna seems, with Al-Gazel, to disagree with most philosophers. For he wrote in 4 *De anima* that there are two natures of soul. Some are elevated, noble, and [capable of] great and marvelous actions, so that they do not operate only in one's own body, with its alterations and changes, but in another [body] and without [any] medium such as the work of the eye in fascinating and causing destruction.[78]

This kind of soul, he says, is capable of throwing a man into a well, or a camel into a bath (*camelum in caladrium*).[79] Avicenna, he says, thinks that an inferior nature will obey a superior soul, and thus "such a soul can heal the sick, weaken the healthy, corrupt natures, and change elements, making what is not fire into fire or what is not earth, earth".[80] In all this he follows Avicenna almost verbatim.[81] For Avicenna, this elevated soul was associated with prophecy and miracles.[82] Belief in this kind of psychic power was largely seen as a threat to divine omnipotence and had been condemned in 1277, with the example of the camel being thrown into a bath specifically mentioned.[83]

This powerful soul differs from normal souls, which "correspond more with the opinion of Aristotle and the Commentator [Averroes]".[84] Normal souls, according to Avicenna's text, require appropriate tools to act on matter, "like a bench-maker, who has the form of a bench in [his mind], but is unable to produce [a bench] except through instruments and media, because of weakness and heaviness of his soul".[85] This weaker soul certainly agrees more with what Pietro says elsewhere, where he continually em-

78 *Conciliator*, diff. 135, ed. Boughan 2006: 296–297: "Propter huius secundum sciendum quod Avicenna, visus cum Algazele in parte, a magno philosophorum discrepare. Scripsit namque in quarto *De anima*, quod anime duplicis naturae. Quedam enim sic sunt elate, nobiles, et tam grandium et mirabilium operationum, ut non solum operentur in corpore proprio, factis alterationibus et transmutationibus, verum etiam alieno, et absque medio, quale opus oculi fascinantis et exterminationem operantis existit. Vult enim hinc hominem deiicere in puteum et camelum in caldarium."

79 The camel example comes from Al-Ghazzeli, *Metaphysica*, ed. Muckle: 1933: 193. Cited in Boughan 2006: 297.

80 *Conciliator*, diff. 135, ed. Boughan 2006: 297–298: "Sentit enim quod natura inferior tanquam nobilioris obediat nutu superioris. Et ideo anima talis infirmos samare potest, et sanos debilitare, pravare naturas, et mutare elementa, adeo ut quod non est ignis ignis efficitur, et non terra, terra."

81 Avicenna, *De anima* 4.4 (ed. Van Riet 1968: 65–66).

82 Hasse 2000: 154–174, and Karnes 2015: 355–357.

83 "Quod intelligentiae superiors imprimunt in inferiors sicut anima una imprimit in aliam, et per talem impressionem incantatory alicuis proiecit camelum in foveam solo visu." (Item 63 in Hissette 1977: 116).

84 *Conciliator*, diff. 135, ed. Boughan 2006: 301: "Hic itaque modus convenit magis cum sententia Aristotelis et Commentatoris ponentium nullam operationem ab aliquo provenire sine motu et transmutatione in hec inferiora. Immo omne agens, quantumcunque fuerit magis distans a contactu patientis, pluribus instrumentis et mediis indigere, ut contrahatur et dearticularetur ad operandum."

85 *Conciliator*, diff. 135, ed. Boughan 2006: 299: "Est autem aliter anima, debilis, infirma, non valens in operationem sine instrumentis et mediis devenire, velut apparet in scamnificante qui formam scamni in ipsum introducere non valet nisi per instrumenta et media, propter debilitatem et otiositatem anime ipsius."

phasizes the need for the appropriate disposition and interaction of both the physician and the patient.

Opinions have differed about how to interpret Pietro's extremely undigested summary of Avicenna's position. Thorndike simply lists it among the many phenomena endorsed by Pietro.[86] Delaurenti considers it more carefully, and particularly emphasizes the role of the enchanter's imagination in all of the effects produced by spells, regarding the elevated soul, freed from worldly pleasures, in possession of great power, and able to act at a distance, to be the pinnacle of the soul's powers proposed by Pietro.[87] On the other hand, Robert, while rightly emphasizing the causal chain between the souls of the doctor and patient in the normal course of events, also argues that Pietro, in accord with the rest of his medical contemporaries, clearly rejects the Avicennan notion of an elevated soul.[88]

In fact, it is difficult to say where Pietro stands. He is clear that the theologians (*Christocoli*) disagree with this kind of powerful soul and that Aristotelians find it ridiculous (*ridiculosa*), but he never says in his own words that he rejects the Avicennan position, as he does with the power of words.[89] Thus it is a stretch to say that Pietro is unequivocal in his rejection of the elevated soul. After all, at other times, when he wants to clearly show his disapproval, he does not hesitate to do so! Further, when we look at the causes he mentions later in his discussions of spells, the first of them is the soul that is so elevated that it is able to dominate matter.[90] Does he have Avicenna in mind here? He does not say. In the end, it is likely (but not certain) that he does have at least some sympathy for Avicenna's position. Maybe he hopes it could be true. However, perhaps because it had been specifically condemned and was almost universally rejected by his colleagues, he does not endorse it, but only mentions it by way of suggestion.

What is clear, is that in these two influential *differentiae*, Pietro moves from the commonplace to the extraordinary, from phenomena with which Galen and Aristotle would clearly agree, to others with which they certainly would not, characteristically flattening the differences between them. In the process, he articulates the most complete description of the interaction of the levels of soul to be found in his work.

**86** Thorndike 1923: 902

**87** Delaurenti 2007: 135.

**88** Robert 2014: 482.

**89** *Conciliator*, diff. 135, ed. Boughan 2006: 299: "Quorum causam non alicui anime habet Christocoli principaliter attribuere, verum virtuti divine operanti hec miranda supernaturalia per viros tanquam per instrumenta prophetas appellatos, quasi medium quoddam inter nos et Deum obtinentes. Hec autem non parum, ut apparebit opinio Peripateticis facta, est ridiculosa."

**90** *Conciliator*, diff. 156, ed. Delaurenti 2012: 82: "Aut ex parte incantantis affectuosi, cum eius fuerit anima in tantum elata ut materie possit huic dominari mundane, ipsamque vigore intelligentie suscepto ad nutum transmutatre."

# Chapter 6
# The Physiology of Moral Virtue

In *differentia* 3 of the *Conciliator*, Pietro comments that the difference between the way that medicine and moral philosophy approach the soul's improvement is that moral philosophy attempts to teach moral doctrine directly, when in fact virtue is acquired by the slow development of natural virtue and habituation of the body through actions.[1] He concludes this thought with a maxim he attributes to Galen that "whoever wishes to heal the soul, should first heal the body", backing it up with a similar sentiment he attributes to Aristotle. Exactly what he means by this is most clear in his treatment of a set of particular moral problems in his *Expositio Problematum Aristotelis*, his commentary on a collection of pseudo-Aristotelian problems.[2] Without presenting a systematic moral theory or psychology, his glosses on this eclectic assembly of questions nevertheless provide a picture of the way physiology interacts with moral psychology in Pietro's thought, while at the same time offering us broader insight into his anthropology. What is apparent in this commentary is that not only the lower virtues, but also those dependent on the highest rational faculties are closely tied to physiological states. Moral virtues and vices closely coordinate with the body's physical balance. He brings a distinctively medical perspective to these moral problems, reevaluating dominant philosophical or theological approaches to virtue with an approach that is at once more material and more attentive to the need for practical intervention. As with so many other aspects of his thought, here Pietro exhibits his tendency to propose medical philosophy as an alternative perspective on philosophical issues of his time.

## Aristotle and the Philosophical Import of the *Problemata*

Pietro's arguments about the philosophical significance of the material found in the *Problemata* follow from his defense of its Aristotelian authorship.[3] Because of the authorship, he dismisses the notion that the *Problemata* is merely a medical text, saying

1 *Conciliator*, diff 3, (f. 7, 3F–G): "in [moralia] doceantur mores et virtutes directe naturales. Acquisiti vero remote ac extraneae valde. Iuxta illud, Primo *De Ingenio*, cap. 5: Qui vult curare animam primo curet corpus. Et *Politicorum*, 7: Corporis quidam curam necessarium priorem fore quam quae anime."

This chapter is a revised version of my article, which orginally appeared in *Early Science and Medicine* (Klemm 2006). I thank Brill for permission to reproduce it here.

2 For further description of the *Problemata* and Pietro's Commentary, see the Introduction: 11–13.

3 *Expositio*, Prologue (ed. De Leemans 2016: 47–48): "Efficiens siquidem huius scientie fuit Aristotiles Steyritanus, filius Nicomachi, sapientissimus Grecorum, qui secundum Averoym dignius meretur dici divinus quam humanus. Neque est quod hic liber non fuerit Aristotilis sicut asserunt quidam mendaciter abnegantes, cum de ipso in secundo Metheorum ac in plerisque locis Parvorum Naturalium et libris Animalium faciat multotiens mentionem ac in hiis quidem derelicta hoc opere compleantur."

https://doi.org/10.1515/9783110612639-007

that some (unnamed) people employ the medical categories of naturals, non-naturals, and contra-naturals to divide its 38 sections (or *particulae*).[4] This medical classification, he replies, is inappropriate because Aristotle does not use the term "natural" in such a way.[5] Nor does it sufficiently reflect the depth of its philosophical material, which is a compilation of discourses on every branch of philosophy, with the exception of grammar and metaphysics.[6] Pietro proposes an alternative three-fold division:

1) medical science and the physical nature of man, sections 1–14;
2) the seven liberal arts, sections 15–19; and
3) problems about natural entities both composite and simple in sections 20–38, which also include the questions about *moralia* in sections 27–30.[7]

Pietro argues that this structure, in particular the decision to begin with medical questions, is purposeful and reflects epistemological assumptions evident in Aristotle's other works:[8]

> Aristotle began by treating questions on medicine for two reasons: [1] Because what is more necessary for us is [more] enjoyable and takes precedence [over what is less needed], as [Aristotle] says himself in the beginning of the Metaphysics: we work harder for what is more necessary. And since we enjoy health more than many other things, in working on this book, he came to start with a discussion of medicine, of which the principal end is health. [2] Because medical things are more known to us, since they are treated in the science which is the knowledge of the human [organism] as it participates in health. For each person ought to know himself better than other things, and therefore, as from what is more known, in the manner treated in the beginning of the *Physics*, he begins to treat medical problems.[9]

---

4 *Expositio*, Prologue (ed. De Leemans 2016: 51): "Dividitur autem hic liber in tres partes secundum quosdam: primo enim determinat problemata rerum existentium contra naturam, secundo naturalium, tertio eorum que sunt non naturalia." As it is used in this classification, "natural" is used narrowly to mean something innate to the human body; thus, everything outside of the body is a "non-natural" or "contra-natural", however natural it might be from a philosophical perspective. On this division see Chapter 1: 19–21.

5 *Expositio*, Prologue (ed. De Leemans 2016: 51–52): "Amplius verba medicorum proponentes non videntur eadem intelligere; non enim quod Aristotiles vocat particulam naturalium, utpote decimam, sic naturale accipit ut medici de re naturali loquuntur, sicut apparet intelligenti sermones eorum et que dicuntur de natura in secundo Physicorum."

6 *Expositio*, Prologue (ed. De Leemans 2016: 49): "Dico autem 'compilationis speciem,' id est modum, quia in hoc libro inveniuntur fere totius philosophie per modum cuiusdam colligationis sermones compilati."

7 *Expositio*, Prologue (ed. De Leemans 2016: 52): "Propter quod dico quod liber hic in tres partes secatur; in prima quarum agit de problematibus circa scientiam medicinalem existentibus et naturalem prout ad hominis materiam reducuntur, secundo de problematibus circa septem artes dictas liberales, tertio agit de problematibus circa naturalia tam composita universaliter et in specie quam simplicia et similiter moralia."

8 This epistemology more accurately reflects Pietro's own understanding of medical science, which he describes in the *Conciliator*, diff. 3 (ff. 5v–8r) and 8 (ff. 11r–14v).

9 *Expositio*, Prologue: "Scias quoque incepisse Aristotelem tractare questiones de medicina duplici ratione: [1] tum quia ipsa est nobis magis necessaria et sic dilecta et preposita – sicut apparet per ipsum

Pietro's discovery of a coherent structure contrasts with the impressions of most modern observers who remark the eclectic, disorderly and unsystematic nature of the text.[10] The defense of the order and authoritative origin of the text has the effect of both enhancing its philosophical value and signaling that the problems will not be treated as strange and wondrous phenomena, but will be reduced to principles. Indeed, however bizarre any problems might be, Pietro is ingenious in attempting to explain just about anything, imposing order through an array of interpretive tools.

With the logic of the text established, Pietro defines the content of each section with reference to the Aristotelian corpus. For this reason, the vocabulary of virtue employed in the *Expositio* is familiar from other scholastic literature inspired by Aristotle's ethical and psychological works. "Virtues", according to this terminology, are habits of regulating our actions and passions (or emotions) in an excellent manner, defined as the mean between two vices marked by excess and insufficiency. Through the possession of virtuous habits we are able to overcome the passions that inspire vice and achieve happiness, the final goal of ethics. The habits are divided into the irrational and rational, following this division in the soul, but must all correspond in some way to a rational principle to be truly virtuous. In the *Problemata*, sections 27–30 progress from the lower to the higher virtues. The irrational virtues are represented by the common division between the "irascible" (aggressive) habit of courage in section 27 and the "concupiscible" (desiring) habits of continence and temperance in section 28. These are followed by a section about justice and the rational habits, including prudence, intellect, and wisdom.

Despite the familiarity of this moral vocabulary, the questions in the *Problemata* are very different from those that preoccupied the faculties of arts and theology at Paris.[11] In fact, without the benefit of Pietro's arguments, it would not be obvious that they are about moral philosophy at all. The dominant issues of the time, notably the debate about the primacy of the will or intellect in motivation, are hardly represented in the questions.[12] There is rarely a distinction made between actions that have "natural" or physiological causes, actions that are completed out of habit, and those that

in principio *Metaphysice*, necessarioribus quidem citius insudamus – et ita cum sanitatem pre aliis multis diligamus in hoc libro tractatis, fecit precedere sermonem de medicina, cuius finis principalis est sanitas, [2] tum quoniam medicinalia sunt nobis notiora, cum tractentur in scientia que est hominis cognitio ut sanitate participet. Unumquodque enim melius se quam alia debet cognoscere et ideo, tamquam a notioribus, secundum modum traditum in principio *Physicorum*, incepit problemata medicinalia tractare."

**10** E.g., see the remarks in Blair 1999: 174–175, and Cadden 2013: 8–9.

**11** The situation in the Italian universities was different because there were no theology faculties and there were no *Ethics* commentaries at this time. See Lines 2002.

**12** On the centrality of this debate in ethical discussion at Pietro's time see Kent 1995.

are rationally chosen and willed.[13] In short, the role of "intention" (in the modern sense), a matter of considerable importance to Pietro's contemporaries, is muted in the naturalistic approach of the text and commentary. For the medieval commentators on Aristotle, that an action is in some way willed defines it as moral rather than natural.[14] Yet, this generic distinction is rarely acknowledged in Pietro's *Expositio*; only the final goal of achieving happiness through the perfection of the virtues remains intact. It is difficult to characterize the problems from these sections, but their starting point is typically the physical signs and symptoms which accompany various types of virtuous and vicious emotion and behavior. The analysis of these questions in the Aristotelian text is generally not what we would call moral or psychological, but rather implicates material and efficient causes in the body. Because of this characteristic of the *Problemata*, common to the genre of problem literature as a whole, it can justifiably be asked whether Pietro is simply following the spirit of his text or whether we are actually hearing Pietro's voice. While this question can be asked of any commentary, in this case it can be satisfactorily answered by close attention to the text. Pietro's rich descriptions of every element of each question offer many opportunities to gauge his opinion. The frequent doubts, digressions, and clarifications are particularly useful for identifying Pietro's own point of view. Through these and more subtle means, he does not hesitate to steer the interpretation in a specific direction. For this reason, even when Pietro simply echoes or minimally amplifies the position of the text (as is frequently the case in the examples from *particula* 27), it is fair to assume that he generally agrees with the position. The numerous instances where Pietro misunderstands difficult passages are instructive as well, because they often reveal his assumptions and expectations about the meaning of the text.[15]

Even if one considers the physiological bent of the *Problemata* and the numerous available *loci* within the Aristotelian corpus for the physiology of human motivation, Pietro's approach to moral virtue in the *Expositio* is unusually attentive to the material conditions necessary for virtuous behavior and reliant on medical doctrine. Moral psychology is examined as it is manifested in the physical state of the body and governed by physiological change. Pietro's insistence that these are moral questions signals that this is a very different approach to human virtue than that of his contemporaries, blurring the distinction between moral virtue and natural virtue throughout the commentary.

---

13 See Cadden 2013: 35–105, especially 69–72, on the conflation of these causes (natural, habitual, and mental), in Pietro's analysis of sodomy.

14 On freedom of the will and morality, see Pink 2012. For Aristotle's discussion of voluntary action, see *Nichomachean Ethics* III.1 = 1109b30–1111b4 (ed. and transl. Rackham 1926: 117–129).

15 There are several occasions in these problems where Pietro is led astray by corruptions in the manuscript tradition, a faulty translation, or unfamiliarity with the subject matter. See Coucke and Swaenepoel 2009.

## *Particula* 27: Irascible Virtue

"Why do the fearful tremble? Is it because of cooling? For their heat is lacking and is contracted, so that in many people the bowels are loosened."[16] These rather ignoble symptoms of fear which open this section about the "irascible virtue" of courage (*fortitudo*) are apt representatives of the material here, in which such symptoms of fear (*timor*) are the subject of seven of the eleven questions.[17] To the trembling and loose bowels of the fearful, other symptoms are added in subsequent problems: people in a state of fear are sometimes thirsty (27.2); their pulse is weak and sputtering (27.3); their faces turn pale (27.3); their voices quiver, and their lower lip and hands tremble (27.6 and 27.7); they emit choler (bile) in their urine and their testicles contract (27.7 and 27.11) and the truly terrified may even emit semen (27.11). The courageous, on the other hand, frequently are lovers of wine (27.4), who are also sometimes extremely thirsty and tremble, but for reasons different from the fearful (27.3).

This list of the symptoms of fear and courage function as the signs which point to their natural causes, which are provided by the physiological theory of the body's complexion – the balance of the four primary qualities of hot, cold, wet, and dry.[18] Pietro's explanations, consistent with those in the Aristotelian text, rely on the fluctuation in the intensity of these qualities. Further elaboration is supplied by the movement of the humors and spirits through which the qualities are transferred. Based on this physiology, fear is the result of the movement of heat from the exterior parts of the body to the interior; recklessness (*audacia*) is caused by the opposite movement, as the heat expands outward from the heart to the exterior of the body.[19] Courage is also the result of the expansion of heat, but presumably to a lesser extent. Pietro refines these general definitions in the explanation of each symptom.

---

16 *Problemata*, 27.1: "Propter quid timentes tremunt? Aut propter infrigidationem? Deficit enim calidum et contrahitur; propter quod et ventres solvuntur multis."

17 *Expositio*, 27.1: "Determinat igitur primo problemata circa habitum moralem, qui est fortitudo vel audacia et eius oppositum, puta timor. Fortitudo enim virtus est, illa vero vitium (*Ethicorum* tertio) . . . Est autem fortitudo, de qua sermo, virtus per quam activi sunt bonorum operum in periculis ut lex precipit et legi obsequentes, *Rhetoricorum* primo. Audacia vero et timor contrarium. Ipse tamen id notat motu interveniente de partibus extrinsecis ad intrinseca sicut audacia magis econtra. Vel fortitudo est virtus irascibilis non facile obstupescibilis a timoribus qui circa mortem *De Bonis Aristotelis Laudabilibus*."

18 See Chapter 2 for my discussion of complexion.

19 *Expositio*, 27.1: "Solvit, dicens causam [timoris] esse propter infrigidationem partium que sunt exterius, calido ab eis separato et ad inferiora intrinseca confestim contracta." And again: 27.1 "Audacia vero et timor contrarium. Ipse tamen id notat motu interveniente de partibus extrinsecis ad intrinseca sicut audacia magis econtra."

The trembling of *timentes* occurs mostly in the extremities and the surface of the body because the heat abandons these parts first, leaving the afflicted person unable to control the affected parts, as when someone strains to raise (*sublevare*) his hands while suffering a "bad disposition".[20] This trembling happens first and most severely in the lower lip and hands because they have the least amount of blood of all the parts of the body (with the exception of the eyes).[21] The paleness in the face and trembling of the voice are caused by the loss of heat, spirits, and blood from the upper parts of the body's surface, extremities and chest.[22] The involuntary loosening of the bowels indicates that the sudden contraction of heat can be more precisely located in the lower part of the torso, heating and loosening that area.[23] The thirst of the fearful is also explained by the location of the heat inside: despite the exterior cooling, part of the interior is hot, thus the sensation of thirst arises, even if it is not properly thirst.[24] The thirst of brave people however is much stronger. This is because the heat is not only much greater in the brave, but the heat collects further up in the torso around the heart, causing the blood and spirits to boil up.[25] The difference in the pulse in anger and in fear also indicates the activity of the heat. In fear, the pulse

20 *Expositio*, 27.1: "Proponit igitur primo, dicens quare timentes membra sua non possunt regere et quietare, sed passionem incurrunt que dicitur tremor, in quo est motus "non naturalis" causatus ex natura membrorum conanti sublevare et prava dispositione ipsius subprimere."

21 *Expositio*, 27.7: "Assignat causam quare labium et manus contremescunt dicens id esse quia labium inferius et manus sunt mobilissima post oculos, cum sint musculosiora, quasi nihil participant sanguine respectu aliarum partium interiorum, quo quis defenditur a rigore propter calidum. Quantum enim membrum de sanguine tantum habet de calido (*Complexionum* primo); merito igitur rigescunt." It is only the lower lip that trembles, because it relies more on heat and spirits (which is why it points up, against gravity). Pietro helpfully cautions us that this trembling is also a sign of impending vomit.

22 *Expositio*, 27.6: "Solvit, dans primo causam universaliter, dicens quia timor est passio quedam, in qua est defectus calidi, quod erat in partibus superioribus ut capite et pectore, deorsum tendens etiam sub diafragmate non parum. Et facies et universaliter partes superiores deserte sanguine et spiritu redduntur pallide."

23 *Expositio*, 27.1: "Calidum habens in exterioribus membra regere deficit. Propter cuius quidem subitam inferius intrinsecam contractionem contingit ulterius ventres solvi ac urinas etiam provocari *absque electione multociens.*" (my italic)

24 *Expositio*, 27.2: "Solvit primo quasi distinctione, dicens ad motivum quod frigiditas et caliditas in eis non sunt secundum eundem locum sed diversum, quia frigiditas est in superficie. Caliditas vero circa interiora conculcata quare calefacit in loco ubi est." 27.3: ". . . non proprie sitis dicens quod licet timentibus accidat aliqualis esiccatio superficialis sanguine confugiente introrsum. Non tamen dicendum proprie sitis."

25 *Expositio*, 27.4: "Solvit primo assignans causam quesiti, dicens id esse quia illi qui sunt fortes animosi sunt calidi. Huiusmodi autem caliditas existit maxime circa pectus in concavitate cordis in quibus sanguis ebullit et spiritus. Circa quem locum videtur timor adesse, cum infrigidatio quedam occurerit. Quare in timentibus calidum minus permanet circa cor. Ymo infrigidatio ad ipsum ascendit cum in timore calidum non solum coadunetur interius in ipso corde, ymo inferius tendat magis sub diafragmate, ut apparebit in sequenti. Sic itaque tantum habitum, quod fortes animosi caliditatem habent vehementem circa cor."

sputters like a fire lacking ventilation while in anger the heat around the heart is amply ventilated and increases.[26]

*Problem* 27.3 compares anger (*ira*) and joy (*gaudium*) to fear and courage, further clarifying the irascible movement of heat in the body, asking why, when the heat collects inside in angry people, they become more hot and excessively bold (*audaces*), but the opposite happens in fear.[27] Referring to Galen (*De accidenti et morbo*), Pietro first clarifies the difference between these passions, noting that it is not exactly correct to say that the heat and moisture are collected inside in anger; rather, this is partly true, for at the same time the heat is sent outward. The outward tendency of anger is manifested by an appetite for vengeance; meanwhile the heat retained inside around the heart acts as a natural defence mechanism, protecting the heart.[28] For this reason, he notes, nobody dies from an overabundance of anger, as they can from fear and joy when the heat abandons the heart – in fear it moves lower and in joy it gets dispersed everywhere.[29] The differences in the movement of the heat defines each of these passions.

These physiological explanations do not seem particularly "moral", especially since most of these passions and resulting actions are done without choice, as Pietro often points out. The essentially physiological and irrational causes of these actions are emphasized in *Problem* 27.10, which compares the irascible movement of heat in the body to the movement of an animal, acting solely by an impulse to pursue or flee and without reason.[30] The involuntary nature of this behavior is a point of continual

---

**26** *Expositio*, 27.3: ". . . in timentibus propter debilitatem virtutis redditur saltus spissus et pungitivus propter eius subtilitatem – subtile namque pungit – vel calor efficitur dempsus et pungitivus propter nimiam humorum et maxime colericorum ebulitionem ob nimiam ipsius congregationem, ceu ignis carens respiraculo. Sed in aliis, ut in iratis, oppositum contingit. Saltus enim sit cordis propter collectionem et debitam aggregationem intrinsecus multi calidi in quantitate cum eventatione sufficienti."

**27** *Expositio*, 27.3: "Quare cum humidum calefactum coadcervetur interius, irati redduntur calidi et audaces, sed in timore contrarium contingit. Efficiuntur enim frigidi et paventes, cum tamen humidum congregetur interius cum calido in eo [timore] sicut in ira."

**28** *Expositio*, 27.3: "Dicendum quod in ira calidum et humidum partim emittitur extra et partim retinetur. Emittitur quidem ut appareat appetitus vindicte, secundum vero partem retinetur circa cor ut id tanquam a re defendatur summa si expediat. Scit enim natura rebellionem habere foras quare conservat sibi calidum defensivum multum circa cor."

**29** *Expositio*, 27.3: "Et ideo nunquam aliquis mortuus est ex ira (quinto, *De Accidenti*). In timore vero calor tantum emittitur intra propter deiectionem quam timet ab exterioribus omnifariam. Sed in gaudio totaliter foras. Natura enim tota se ad extra expandit cum non sentiat impugnatorem et plures iam mortui sunt pusilanimes ex timore ac gaudio. Ibidem." He adds later in the same problem that inspiring anger in people who are near death because of cold is a good way to revive them, according to Haly Ridwan: "Et ideo possunt seipsos reviviscere cum fuerint infrigidati et quasi extinctioni propinqui, et per consequens concitari vehementer in iram et perturbati effici. Sicut recitat Haly in *Tegni*."

**30** *Expositio*, 27.10: "Dicit id esse quia calidum innatum quod nobis inexistit est sicut animal quoddam, ut enim ipsum fugit ab eo quod timet et abscondit se et accedens ad id unde sperat iuvari, similiter et calidum animalis; unde videmus ipsum sub ratione iuvativi aut nocitivi moveri ad exteriora

emphasis in Pietro's commentary, even when it is not mentioned in the base text. In 27.9, the question asks why it is that the fearful are silent while those in pain cry out, since fear seems to be a type of pain.[31] The difference results from the fact that the people who are suffering hold their breath and then release it suddenly, but in fear the heat and spirits leave the organs necessary for speech. This is an involuntary reaction that builds up heat in the body to relieve the suffering of pain, which he compares to the natural instincts of other animals.[32]

At no point in the commentary to this *particula* does he make a moral distinction between considered choices of the soul and simple physical processes as a source of motivation. Complexion is not only a sufficient cause of individual actions, but it is also responsible for the habits which are more important to the constitution of moral virtue or vice. Pietro takes the opportunity in his commentary to *Problem* 27.4, which asks why courageous people love wine, to explain the transition from qualitative imbalance to inclination to habit, describing how internal heat forms the habits of both heavy wine-drinking and bravery. Heat around the heart causes the blood and spirits to boil. When this heat is so great that it remains uncooled even in the presence of great danger, it causes people to be both courageous and lovers of wine. The constant need to be cooled produced by this heat causes a desire for wine which eventually becomes a habit (love of wine) when the desire is repeatedly indulged. Wine, being a warming drink, only increases the desire for more. The appetite caused by the hot complexion leads to the love of wine and simultaneously contributes to courage because: "it is often necessary for natures and complexions to follow each other, so that being brave and the desire for cooling follow from heat".[33] If the heat should stop, appetite and inclination can be reversed.[34]

Again in *Problem* 27.3, Pietro expands the remarks in his text on the role of heat in producing bravery.[35] This problem describes the tendency of some brave soldiers

---

modo, tunc autem interius, cui matrix assimilatur *Interiorum* et Serapionis sextis auctoritate Platonis ita ut extimaverit eam fore animal eo quod sic movetur ad odorifera et refugit opposita. Quia igitur, ut dictum est, calidum ita refugit ut animal nocivum frigidis utique ut exterioribus nocentibus et id aut propter pugnam vel timores vel aliqua alia sintomata calida vel anime, calidum refugit et abscondit se de partibus superioribus ad interiores et ex locis superficialibus ad profunda."

31 *Expositio*, 27.9: "Quare cum timor sit ut quedam tristicia et dolor similiter (et id est motivum) ipsius qui dolent clamant et vociferant, sed qui timent silent."

32 *Expositio*, 27.9: "Causa extat quia omnes qui dolemus etiam *sine electione* utimur subsidiis que naturaliter sunt nobis inserta." (my italic).

33 *Expositio*, 27.4: "Amatores igitur erunt potationum vinosarum, quia necessarium est multociens naturas et complexiones consequi se adinvicem, ut ad esse forte sequitur calidum et appetitus infrigidationis."

34 *Expositio*, 27.4: "Cum autem vehemens caliditas cessaverit appetitus etiam et inclinatio quam habuerunt ad vinum cessat."

35 Pietro did not completely understand this text because of corruptions in the text manuscript(s) he used. E.g. "in bellis" (in warfare) has been rendered as "imbecellis" in the *Problemata*.

to beat themselves with fennel(!) before battle to stop their trembling, explaining that this is not because they are afraid, but that their courage and the impending battle have created a "disturbing inequality" in their bodies' complexions. Misunderstanding the text slightly, Pietro assumes that the soldiers are afraid, but not so much that they are stupefied, so they bludgeon themselves to heat the blood and spirits in their bodies and thereby make themselves brave. Far from being a distinctively human reaction, he notes that this is similar to lions, who use their tails for the same purpose.[36]

Although they are not mentioned by Pietro, it is interesting to compare Aristotle's comments in *Ethica Nicomachea*, 3 to the effect that courage caused by physiology cannot truly be called the moral virtue of courage.[37] But Pietro does not make a moral distinction between physiology and rational choice as causal agents in this *particula*, nor is it clear that he would ever make such a distinction. Complexion for Pietro is both the immediate potential for action shared by the body and soul and the instrument through which action is brought about. Courage appears in this commentary as a kind of habituation of the body by maintenance of the qualities, specifically a healthy amount of heat, presumably produced more by regimen and physical manipulation than by rational choice and deliberation.

## *Particula* 28: Concupiscible Virtue

The short *Particula* 28 continues the virtues of the sensitive soul with eight questions on the "concupiscible" virtues and vices of "temperance, intemperance, continence, and incontinence",[38] where again the physiological causes of desire are prominent. Pietro provides the standard Aristotelian definitions of these moral terms, clarifying that continence and temperance are distinct habits, but that only temperance is a true virtue, a definition that ignores the condemnation of 1277, when this meaning of continence was censured.[39] Continence falls short of virtue because a continent person still suffers the

**36** *Expositio*, 27.3: "Licet timorem aliquem incurrant, non tamen stupefiunt et sensibus deficientes redduntur. Ymo illi qui eorum sunt fortiores, solent se percutere prius antequam accedant ad pugnam, membra sua ferula ampla valde, ut multitudinem caloris suscitent intus, quatenus amplius fortiores reddantur et animosiores in pugna, ceu dicitur de leone se cauda stimulante. Quid autem sit ferula visum nona. Quod si non se ferula flagellent, fricant se manibus per totum ut calefiant et sanguis et spiritus diffundatur per omne non remanens contractus interius omnino."

**37** Aristotle, *Ethica Nicomachea* 3.8 = 1116b24–1117a28 (ed. and transl. Rackham 1926: 167–171).

**38** *Expositio*, 28.1: "Egit de problematibus circa irascibilem virtutem, nunc circa concupiscibilem in hac 28[a] circa quam consistit continentia et incontinentia temperantia et intemperantia."

**39** "Quod continentia non est essentialiter virtus", is item 168 in the list of banned doctrines in the *Chartularium Universitatis Parisiensis*, ed. Denifle and Chatelain 1889–1891: 1.543–555. Aquinas also used continence in this manner prior to 1277, e.g. *Summa Theologiae* Part 2–2, q. 155. art. 1, resp (transl. Freddoso: 972). For Aristotle's definitions see *Ethica Nicomachea* 7, 1146a10–16 in particular (ed. and transl. Rackham 1926: 381).

passion of desire for illicit pleasures, but resists;[40] a temperate person, however, is so constituted that such pleasures have become undesirable.[41] Incontinence and intemperance mirror these definitions: for the incontinent, reason has forbidden illicit behavior, but he is unable to resist passion – once the weakness subsides he feels remorse and there is still hope for such a person to be cured.[42] The intemperate is almost completely beyond cure because he is fully directed toward the wrong actions.[43] The five external senses are the main focus of these questions because the corporeal desires governed by concupiscible habits are sensual, especially sensual pleasures related to touch and taste, which is why we call people incontinent and intemperate especially in regard to these two senses.[44] Pietro's tone is more judgmental than in the previous section, repeatedly indicating his contempt for the indiscretions of touch and taste as well as his insistence on the importance of sensual virtue for human perfection.[45] The sensitive soul, when properly

---

**40** *Expositio*, 28.1: "Continentia vero est habitus concupiscibilis, secundum quem retinet ratione concupiscentiam moventem ad pravas fruitiones delectationum."

**41** *Expositio*, 28.1: ". . . temperantia sit virtus concupiscibilis secundum quam inappetibiles fiunt circa fruitiones pravarum delectationum."

**42** *Expositio*, 28.1: "Incontinentia vero est malitia concupiscibilis secundum quam eligunt pravas delectationes prohibente rationali. Cuius siquidem opera prohibente rationali fruitiones delectabilium eligere et suspicare melius esse non participare ipsis, participare autem nihilominus scire oportere operari bona et conferentia, recedere autem propter delectationes ab ipsis. Consequitur autem incontinentie mollities, penitudo et sanatio."

**43** *Expositio*, 28.1: "Intemperantia vero est malitia concupiscibilis, secundum quam electivi fiunt earum que circa voluptates delectationum pravarum. Cuius quidem opera sunt eligere fruitiones nocitiarum et turpium delectationum et suspicari felices fore maxime in talibus delectationibus viventes et amativum risus esse convitiorum et eutropelie et fallacem seu levem in sermonibus et operibus. Ei autem consequitur inordinatio, inverecundia, indecentia, delicie, desidia, negligentia, incuria et dissolutio." And ibid: "Sed intemperatus est qui persequitur superhabundantias delectationum vel secundum quod sunt superhabundantie aut propter electionem et propter ipsas et non propter aliud nihil occurrens. Propter quod sequitur eum non penitere. Unde insanabilis et deterior est incontinente. Hunc autem penitet et sanabilis assimilatur minusque illo malus."

**44** *Expositio*, 28.2: "Quare homines dicuntur incontinentes et intemperati fore maxime secundum hos duos sensus puta tactum et gustum." In addition to the questions about the senses, this section also includes a question about inordinate laughter (28.8), which Pietro misinterprets in a rather humorous manner, and another about why people who are habitually intemperate become ill when they suddenly adopt a more moderate regimen (28.1), which is an opportunity for Pietro to explain the principle that habit becomes a second nature.

**45** E.g., *Expositio*, 28.2: "Et ideo aut soli hii, aut maxime inter ceteros, sunt vituperosi et abhominabiles. Et propterea non inmerito nequeuntes resistere delectationibus, que per eos, sed potius captivati, vincuntur ab eis, vituperamus non parum dicentes: "Quomodo dimittitis vos vinci a tam pravis delectationibus adeo momentaneis." Quod "ut ad multum" ostendit, penitentia et abhominatio subsequentes. Unde incontinentes vocamus et luxuriosos, eo quod vincantur ab huiusmodi delectationibus pravis bestialibus."

The contempt is also registered in 28.3 and 28.7. Cadden 2013: 105 also notes Pietro's disgust at "sodomites" in his natural analysis. The naturalization of sexual practices does not shield them from disapprobation.

ordered and enjoyed, functions as the means to human virtue, but it is through lust (the pleasure of touch) and gluttony (taste) that we are most like beasts, who share our enjoyment of touch and taste alone among the senses.[46] The pleasures of these senses are insidious because they so thoroughly preclude the use of the rational faculties: "Desire concerning [touch and taste] is among other things which greatly obscure and disturb the reason. For the whole man is made bestial, since he is less obedient to reason than even the irascible."[47]

Like the irascible vices, the desires of touch and taste have physical causes, produced by excess and deficiency in the qualities. Thus, thirst is the dual desire for cooling and humidity, caused by the opposite state in the body, and also a desire for nutrition, although desire for food is less powerful.[48] The cause of sexual desire is only mentioned without being fully explained in this *particula* (probably because it is discussed in much greater detail in *particula* 4), when Pietro mentions the tendency of adolescents to be "especially stimulated to sex", because they are most molested by concupiscent passions.[49] It is clear from Pietro's tone throughout this section that these "physiological" explanations do not mitigate Pietro's judgment of the depravity of bestial pleasures.[50]

If excessive pleasure in touch and taste is especially reprehensible, on the other hand appropriate enjoyment of the three higher senses is essential to the development of the distinctively human virtues.[51] Borrowing a hierarchy of the senses from Galen, Pietro explains that touch and taste (even when used virtuously) produce only

**46** *Expositio*, 28.2: "Solvit, dicens causam esse propter delectationes que fiunt in hiis sensibus, nobis et animalibus communes, quantumcumque imperfectis, cum omnibus animalibus necessario hii duo insint sensus."

**47** *Expositio*, 28.3: "Nosce quod intra cetera que magis rationem obnubilant et perturbant est concupiscentia circa pretacta [tactus et gustus]. Totus enim homo brutalis efficitur, cum minus obediat rationi quam etiam irascibilis. Unde septimo *Ethicorum*: Ira consequitur rationem aliqualiter concupiscentia autem minime. Et ideo difficilius est pugnare voluptati quam ire Eraclitus eorundem secundo."

**48** *Expositio*, 28.5: "Et quia sitis est duplex desiderium et indigentia, potus scilicet et cibi, ac fames tantum cibi. Itaque desideramus minus cum propter alterum solum, ut taliter ratiocinetur, illo, quo magis conservantur principia vite pluribusque competit de causis . . . Nosce quod cum potus sit triplex, ut visum est prima, per "potum" auditur ille qui est cibus et potus simul, ut vinum vel aliquis liquor huiusmodi. Nutrit enim, refrigerat, humectat, facit etiam ad distributionem per corpus. Cibus autem nutrit solummodo. Necessarior igitur potus; difficilius itaque abstinetur ab eo." 28.7 specifies the physical location of the desire for food: ". . . delectatio circa cibum, quidam habent eam magis in lingue superiori pellicula, propter multam eius sensibilitatem. Alii vero non solum in ea, sed amplius per totum guttur." Which is further specified as the "spatium cum os aperitur et lingua deorsum plicatur, quod apparet ante foramina epiglotis et isofagi cum parte etiam ipsius aliquali . . ."

**49** *Expositio*, 28.4: "Iuvenes enim a concupiscentiis maxime molestantur . . . Stimulantur namque ad venerea maxime."

**50** Cadden 2001 and 2013.

**51** *Expositio*, 28.7: "Nosce etiam quod potest alia causa dari quare animalium alia sensibus ceteris a tactu et gustu non gaudent nisi per accidens quia cum homo sit sapientissimum animalium et prudentissimum: (*Polliticorum*, primo et secundo, *De Anima*) indiguit instrumentis quibus posset sapientiam

our mere existence (*esse*); smell, hearing and sight, on the other hand, are productive of the virtues which give us well-being (*bene esse*).[52] They should not only be used, but enjoyed, because their enjoyment is fundamental for acquiring prudence and wisdom and the other higher virtues. Animals, Pietro explains, do not take pleasure in sight, smell, or hearing (except accidentally), because they have no rational virtues to develop. As more noble creatures than the other animals, humans ought to order their lives toward higher virtues, since it is especially contemptible for a noble creature to partake in less noble virtues.[53]

Pietro's argument for the necessity of the appropriate sensual habits as the basis for rational virtue has a polemical tone on occasion. Perhaps this tone is to compensate for the fact that this strong thesis about the role of the sensual habits is not explicit in the Aristotelian text, but is only introduced by Pietro in his commentary. It is a theme echoed in the *Conciliator*, where Pietro refers derisively to the notion that human perfection is only achieved in a pure intellectual state "abstracted from the senses".[54] In response to such statements, he frequently cites the Aristotelian maxim that "nothing is in the intellect that was not first in the senses", stressing the dependence of the intellect on the sensitive soul.[55] At the same time, his focus on the training of the base desires as the path to virtue distinguishes Pietro's approach from his contemporaries, especially among those who place moral virtue in the dominance of the will over passion. Our inability to resist base passions combined with the necessity of doing so affirms a role for the physician in moral development, as the higher virtues arise from the physical balance, which is the aim of medicine.[56] This theme becomes more evident in the following sections, where the intellectual virtues are treated. When he turns to justice and

---

et prudentiam exercere. Huiusmodi autem sunt visus, auditus et odoratus, ut apparet *De Sensu et Sensato*."

52 *Expositio*, 28.7: "Nota quod circa tactum contingit potissime delectari, consequenter circa gustum, deinde odoratum, postea auditum, et tandem circa visum minime. Quia tactus ceteris materialior, hic autem spiritualior, medii autem mediocriter se habent (quarto, *De Accidenti*) . . . Hii [visus, auditus et auditus] enim faciunt ad scientiam et ad bene esse. Alii vero duo ad esse simpliciter. Quare homo hiis delectatur, ut possit eorum delectatione sapientiam et prudentiam acquirere cum delectatio propria perficiat operationem (*Ethicorum*, quarto). Et quia animalium reliqua non habent sapientiam et prudentiam hiis non nisi fortasse per accidens delectabuntur." He also quotes a similar statement by Averroes, adding that concupiscible pleasures ". . . are not of the disposition for [man's] being, much less for his perfection." *Expositio*, 28.2: ". . . indignitatem cum sapientes noverint proiecerunt ex seipsis: quia non sunt de dispositionibus facientibus ad esse nedum de perfectionibus (Commentator in accessu *Physici* auditus)."

53 *Expositio*, 28.2: "Sed illud in quo nobilius cum ingnobili convenit est innobile valde et despectum respectu nobilioris quoniam quod nobilius debet ulterius innobilius et perfectius ordinari et non invilius."

54 E.g. see: *Conciliator*, diff. 15 (f. 23r, col. 2, C). See Chapter 1: 43–44.

55 E.g., in the discussion of concept formation in the case of confidence in Chapter 5: 96–100.

56 The effects of regimen, especially that of food and drink, on complexion, and therefore on behavior, are especially clear in the commentary to *Problem* 30.1.

the best complexion for the intellect (discussed in chapter 2), he maintains his physiological reading even when the intent of the base-text strains to support it. Indeed, one of the most notable characteristics of his commentary to the questions on justice is his tendency to misunderstand the text because he is so committed to his point of view.

## *Particula* 29: Justice

Introducing the problems of *Particula* 29, Pietro offers an unusual taxonomy of justice in an effort to incorporate the medical category of "natural justice", the most temperate physical balance possible.[57] He also classifies justice among the virtues of the rational soul, again unusual, since Aristotle had counted it among the moral virtues.[58] As is often the case when Pietro differs from Aristotle, he does not advertise his divergence. To defend his strange division, Pietro is forced to rely on a rather tortured use of authorities. He first gives a common definition from the opening of Justinian's *Institutes* that justice "is a constant and enduring will, distributing its law to each person".[59] "Will", he adds, consists of "right reason", hence its inclusion among the rational virtues.[60] Next, Pietro divides justice into two types: "(quasi)-natural" and "positive", the distinction commonly used for law (*ius*).[61] As evidence for these two types, he is only able to invoke a single passage from Pseudo-Seneca, which vaguely implies a category of "natural" justice.[62] Otherwise, the authorities are silent because these categories were not used for justice. Aristotle also defines two types of justice: a general sense in which it is identical with the excellent use of all other virtue, and a particular sense in which it is a virtue moderating various types of social interaction, further subdivided into distributive and commutative justice.[63] Pietro identifies Aristotle's particular meaning of justice with his own category of "positive" justice, imply-

57 See Chapter 2: 57–60.

58 *Expositio*, 29.1: "Determinatis problematibus, que circa appetitum sensitivum consistunt, nunc magis ea, que circa rationalem, ut iusticiam primo . . . . Secundo circa prudentiam, intellectum et sapientiam sequenti *particula* [30]." Although Pietro seems to back away from this classification at the beginning of *particula* 30. See *Ethica Nicomachea* 5 = 1129a6–11 (ed. and transl. Rackham 1926: 253).

59 *Expositio*, 29.1: "Est namque constans et perpetua voluntas, ius suum unicuique tribuens." Justinian's *Institutes*, ed. Kruger 1872: 1. This definition was commonly quoted without attribution, e.g. Aquinas, *Summa Theologiae*, 2–2, q. 58, art. 1 (transl. Freddoso: 375).

60 *Expositio*, 29.1: "Voluntas vero in ratione propria constat." He is either simply defending his inclusion of justice among the rational virtues, or perhaps he mentions both will and reason to avoid taking a side in this debate.

61 *Expositio*, 29.1: "Videtur autem iustitia duplex: una quidem naturalis quasi, altera vero positiva."

62 *Expositio*, 29.1: "Quod indicat Seneca *de verborum copia*: Iustitia est nature tacita conventio in adiutorium multorum inventa et est divina lex et vinculum societatis humane."

63 *Expositio*, 29.1: "Quinto vero *Ethicorum*, apparet quod iustitia dupliciter potest accipi: vel in communi prout omnem continet fere virtutem. Unde ibidem: iustitia similis est omnis virtus et perfecta maxime virtus, quoniam perfecte virtutis usus est; perfecta quidem, quia habens ipsam non solum ad

ing that Aristotle's general meaning of justice is identical with his own category of "natural", although this is left unstated. These categories are not used by Aristotle or his commentators. Otherwise, Pietro's further sub-divisions of the particular meaning of justice are exactly as they appear in the *Ethica*.[64] Because the general sense of justice encompasses the other virtues, it is also closely related to the concupiscible and irascible virtues. Ideally justice guides these lower capacities, but as we saw in the previous section, it is the passions in the body and lower parts of the soul that most interfere with reason. The effect is that Pietro's approach to justice in this commentary is quite physiological, again focusing on qualitative balance in the body and the appropriate habits of the lower levels of the soul as a prerequisite for possessing the virtue of justice.

Pietro invokes the idea of natural justice to explain *Problem* 29.6, which asks why it is that man is the most unjust of animals despite his education.[65] He observes that justice, taken strictly as a rational power, cannot properly be attributed to other animals, who lack reason; they possess it only by a certain similarity. But using the medical notion of natural justice, it is possible to compare human and animal capacity. He defines this "justice" as the ideal complexion which enables a creature to accomplish all the actions appropriate for itself:[66]

> It is possible to say that the natural justice previously mentioned [in *Problem* 29.1], is in other animals. When they have the quantities and qualities of the elements sufficient for accomplishing their activities, they are said to be "*ad iustitiam*" according to Galen and Haly Abbas. (However, Avicenna only posits the complexion of justice in the human race).[67]

---

se ipsum utitur ea, sed ad alterum. Et propter hoc preclarissima virtutum videtur esse iustitia, et neque Hesperus, neque Lucifer ita admirabilis. Est et alia, propria seu secundum partem, que est virtus propter quam que sua singuli habent ut lex precipit *Rethoricorum*, primo. Que siquidem positiva dicitur, et hec duplex. Una quidem dicitur distributiva, que attenditur in partitione honoris pecunie vel talium secundum quod meruit unusquisque; altera commutativa. Commutationis autem quiddam est voluntarium, puta venditio, emptio, mutatio; aliud involuntarium et istius adhuc quoddam occultum, ut furtum, mechia, veneficium ac falsum testimonium et alterum violentum vel manifestum, ut verberatio, vinculum, mors, rapina, etcetera."

**64** See *Ethica Nicomachea* 5 = 1130b30–1131a9 (ed. and transl. Rackham 1923: 267).

**65** *Expositio*, 29.6 [this is *Problem* 29.7 in early editions of this text as well as modern Greek editions. It should also be noted that the 1475 and 1482 editions have incorrectly matched the texts and commentaries for many of the problems in *particula* 29]: "Propter quid homo participans disciplinam est iniustissimum omnium animalium? Aut plurima mente per communicat maxime igitur delectationes et felicitatem assequitur; hec autem absque iniustitia non sunt."

**66** See Chapter 2 on the *complexio iustitialis*.

**67** *Expositio*, 29.6 [29.7]: "Potest etiam dici in aliis animalibus fore iustitiam naturalem pretactam. Ceu, cum habent de quantitatibus et qualitatibus elementorum, secundum quod merentur ad eorum actus exercendos, dicuntur esse talia 'ad iustitiam' penes Galenum et Haly Abbas. (Nam Avicenna complexionem iustitie solum in specie ponit humana)."

In this way, this "natural" notion of justice helps solve this problem. If justice is conceived as the proper use of every available natural faculty, human justice requires the use of reason. Yet, humans are constantly prevented from the use of their reason by "sensitive and bestial passions". In fact, it is the possession of higher faculties that ensures that we will often be corrupted and "unjust": the mere thought of illicit pleasures cannot be accomplished without injustice – i.e. passions and physical imbalance.[68] As we saw in the previous chapter, thoughts have the ability to produce physical passion: hence, even the thought of something fearsome will make us shiver. At the same time, even though humanity as a genus participates in reason and justice, it is nevertheless unjust "because it does not use them often".[69] Animals, being more likely to perform their proper tasks, are more just by this definition.[70] Pietro's use of the medical notion of the just complexion again blurs the distinction between the virtues of humans and those of animals, and hence the virtues of the rational soul and those of the sensitive soul (also possessed by animals).

One result of Pietro's rigidly physiological approach to justice is that he often misses the distinction between different causes of motivation when it is made in the Aristotelian text. This seems to be the case in *Problem* 29.12, where the text considers why a defendant should have an advantage over an accuser in court. The crucial distinction in the text of the question is between an offence inspired by passion, as many crimes are, and an offence committed with prior intent, such as a false accusation; intent makes the false accusation a greater injustice than the act inspired by passion. Pietro fails to see this as the crucial distinction (in an admittedly difficult text), and instead explains that unjust accusations often arise from passion perverting the reason. Anger, lust, and even love "pervert the truth of human judgment", apparently inspiring poorly judged accusations.[71]

Pietro similarly misses the distinction in the commentary to *Problem* 29.9, another text which he struggles to understand because of the compressed translation. Pietro paraphrases this question as "Why, if a sick person associates with a healthy person, or a weak person with a strong person, or a worse person with a better person, the one is not caused to be like the other; i.e. that person in whom these qualities exist,

---

**68** *Expositio*, 29.6 [29.7]: "Et presertim cum vixerit secundum passiones sensitivas et brutales, appetitus desiderat et studet assequi illecebras delectationes et felicitatem. Talia vero assequi non potest et comprehendere sine iniustitia, cum in hiis recta ratio et voluntas constans corrumpatur."

**69** *Expositio*, 29.6 [29.7]: "Sic itaque licet homo sit participans disciplina et intellectu, quia tamen non utitur hiis sepe, sed oppositis magis, ceteris redditur animalium iniustissimum."

**70** For theories about animal psychic capacities, see De Leemans and Klemm 2009.

**71** *Expositio*, 29.12 [29.13]: "Inconveniens namque et turpe, quod aliquis accusetur iniuste. Quod autem fiat accusatio iniusta, contingere potest, aut propter iram, quia cum iratus quis contra aliquem, pervertitur eius ratio, ut nequeat verum discernere illum satagens destruere omnino; aut propter concupiscentiam, ut apparet in actu venereo; vel propter amorem, quia amor et odium pervertunt hominis iudicii veritatem (in *Centiloquio*)."

namely healthy, strong, and good?"[72] Pietro explains that the vices (sickness, weakness, timidity, and wickedness), as passions, "exist in a state of mutation and change. For they [i.e. the vices] do not quietly possess their subjects, but they always change, until they corrupt them, since their existence is very distant from the mean, which is the root of virtue."[73] The virtues are different, because, being balanced, "they are permanent and inalterable".[74] Pietro fails to recognize the distinction probably intended in the question between the virtues of the body, such as health, which cannot be transferred by social contact, and the virtues of the soul, which can be learned from social interaction with virtuous people. Instead, the distinction he proposes is between the balance and dispassion of the virtuous habits and the mutability and passion of the vices. With this he offers a physical explanation for why the virtues cannot be learned: a balanced habit does not transfer its qualities to another, thus destroying its own balance.[75] Pietro precisely reverses Aristotle's conclusions in his ethical works, and probably intended here, that moral virtues can be taught through socialization. Pietro goes on to question the efficacy of abstract precepts for instilling moral virtue, echoing his thoughts in the *Conciliator* at the beginning of this chapter.[76]

Although Pietro does not distinguish the source of motivation in the case of just actions (reason, habituation, qualities), the existence of physical passion is important for the evaluation of some particular vicious actions, playing a large role in determining moral praise and blame. As in earlier *particulae*, the physical state of the body is the cause of desire; by understanding this physiology, we are in a better position to evaluate unjust action. Thus in *Problem* 28.4 (even if Pietro simply follows his base-text here), the relative moral value of justice and continence is informed by physiol-

72 *Expositio*, 29.9 [29.10]: "Quare si quis infirmantium copuletur sano aut debili forti vel deterior meliori, non propter hoc redditur unusquisque talis, qualis ille cui hec non sunt: puta sanus, fortis et bonus? Quod ostendit, dicens quod habitus acquiritur infirmo, debili vel pravo si coniungatur temperato vel sano iusto sive forti et bono."

73 *Expositio*, 29.9 [29.10]: "Solvit, dicens causam fore quia quedam horum sint opposita predictorum habentium, ut infirmitas, debilitas sive timiditas et malicia, in quadam consistunt mutatione et alteratione. Non enim quiete possident sua substantia, sed semper alterant ut tandem ipsa corrumpant cum esse eorum sit valde a medio elongatum quod est de ratione virtutis."

74 *Expositio*, 29.9 [29.10]: "Sed huius, puta fortitudo aut iustitia, bonitas et sanitas sunt permanentia inalterabilia, et quantum ad animam et quantum ad corpus, secundum quod sunt habitus anime et corporis." His gloss that some of these virtues are habits of the body, some of the soul is intended to show that he is talking about both in this question.

75 *Expositio*, 29.9 [29.10]: "Habitus autem non relinquit suum subiectum, ut aliud afficiat, se destruens."

76 *Expositio*, 29.9 [29.10]: "Et confirmatur quoniam quis non efficitur – puta malus non redditur bonus et timidus fortis etiam inquiescendo a pravis, vel etiam usi ratione studiosa audiendo. Immo se oportet in illis exercere, quoniam ex frequenti bene agere fimus boni, sicut ex frequenti citarizare citariste (secundo, *Ethicorum*). Si enim solum admonitiones audiret, et cum hoc etiam disceret, non operans iam scientiam moralem naturalem constituens quod negatum est. Secundo, *Ethicorum*: non enim eorum negotium est contemplationis gratia ut sciamus quid est virtus, sed ut boni efficiamur."

ogy: youths are praised for being continent, while the old are praised for being just, because the young are more stimulated to sexual pleasure and more in need of being continent.[77] Praise and blame are given according to what desires are molesting us the most and desire is determined by what excesses or defects are present in the body at any point. A more specific example is in 29.12, which asks why a defendant in court wins if the votes are equally divided.[78] Here, Pietro argues that the physical symptoms of fear should be an important consideration in the way justice is rendered in the courtroom; so, the defendant, who is more afraid, is right to have an advantage. Terrified defendants are apt to forget things and not represent themselves properly due to their passion.[79] Commenting further (on a text he again does not understand very well), Pietro remarks that, if someone is accused of a serious crime, it is justifiable if they do not present themselves in court. In this case, fear of torture is an understandable reason not to appear in court.[80] The judge should take the "nature, actions, and habits" into account and consider whether passion has undermined the defendant's ability to reasonably defend himself.[81]

Another example of this is *Problem* 29.13 (yet another problem with which Pietro has some trouble), which asks why the penalty is worse for theft from a public place

---

77 *Expositio*, 28.4: "Iuvenes enim a concupiscentiis maxime molestantur. Et ideo nullus eligit eos duces quia non constat eos esse prudentes (*Topicorum*, tertio). Stimulantur namque ad venerea maxime. Quare Seneca protulit iuvenem castum fore martirem absque sanguinis effusione. Et propter hoc legislatores ut apparebit particula consequenti minus puniunt illos aliis qui causa incitante delinquunt maiori."

**78** *Expositio*, 29.12 [29.13]: "Propter quod, quando et fugienti et prosequenti videntur numeri equales, fugiens vincit?"

**79** *Expositio*, 29.12 [29.13]: "Id declarat in timidis, dicens quod si aliquis consideret in reos timidos multociens inveniet, quod multa derelinquuntur in dictis suis probationibus, que oporteret eos dicere aut oppinari esse pro ipsis, que obmittunt exterminati timore et agonia quam patiuntur, quia rei secundum plurimum semper sunt in maioribus periculis et agonibus valde quam actores. Propter quod, si cum hoc, quod derelinquunt proferre vel exterminare propter timorem in eorum subsidium et cum hoc accusationes eorum adequantur et numeri testium et signa probationibus actoris tunc est manifestum iudicanti, qui ad nullam partium contradictionis debet inclinari, ut apparet primo, *De Anima* et exprimitur Primo Prima [*Canonis*]. Nullum igitur debet latere, quod, si rei non derelinquerent circa probationes ipsorum, obmittentes, que sunt necessaria, quod mox iudicaretur pro eis contra actorem. Non tamen propter hoc, etiam si iudex sciat, quod obmittunt, que deberent dicere, habet omnino pro eis iudicare, quia is debet iudicare secundum allegata, et non secundum propriam cognitionem. Sed loco istius dat dilationes et deliberationes ad considerandum, que habent dicere et respondere contra accusatorem."

**80** *Expositio*, 29.12 [29.13]: "Nosce quod, si aliquis non representat se in iudicio propter timorem tormentorum, carcerum, et huisumodi, ordinavit legislator, ut non in tantum puniatur, sicut si presentati fuerint contra eos, manifeste probatum defendere."

**81** *Expositio*, 29.12 [29.13]: "Debet etiam iudex advertere naturam, actus et mores reorum. Quidam enim sunt timentes et stupefacti adeo, ut nesciant se modo aliquo excusare . . ."

than from a private house.[82] Pietro's discussion of this law is again based on physiology. It seems, he suggests, that the penalty for public theft ought to be less than for private, because the temptation is greater and where there is greater temptation, the crime is more understandable and the punishment should be reduced.[83] For the same reason we do not punish youths as much as old men, who are not so impelled by their passions and are therefore more responsible for their actions.[84] However, the great temptation to steal from a public place is exactly the justification for the death penalty: just as nature balances one excess with another, the desire to steal is quickly cooled by consideration of the harsh penalty.[85] Dispassionate crime is always a greater transgression. This explains why they say God did not forgive Lucifer's sin as He does that of a man or a woman: Lucifer had no external cause to inspire passion.[86]

None of this strays too far from the "Aristotelian" text; nor is the association between such sensitive (irascible and concupiscible) desires and physiology that unusual. The rational virtues present a different challenge. Nevertheless, when he turns to justice and the intellect, he maintains his physiological reading even when the intent of the base-text strains to support it. Indeed, one of the most notable characteristics of his commentary to the questions on justice is his tendency to misunderstand the text because he is so committed to his point of view.

Pietro's *Expositio* displays an approach to the virtues that regards physiology as the primary field of inquiry through which the actions and passions shared by the body and soul are to be investigated and explained. Here, the body's complexion is an efficient cause of actions and habits, a sign of passions, and an indicator of potential activity. For the irrational soul, this close connection with the body is not surprising. However, in the case of the rational soul, the dependence on the body is more unusual and offers an example of Pietro's extension of medical doctrine beyond the usual bounds of medicine. Inherent in his physiological approach, is his refusal to regard

---

**82** *Expositio*, 29.13 [29.14]: "Quare, si contingerit quandoque aliquem furari in balneo, ut naturali aut stupha, vel in pallestra aut in foro vel in ecclesia, et universaliter in loco communi ubi homines putant sua esse salva, et ideo exponunt, sicut balneantes et mercantes, punitur ad mortem si devenerit ad manus iudicis. Sed ille, qui furatur in domo, et universaliter in loco privato, non privatur vita?"

**83** *Expositio*; 29.13 [29.14]: "Videtur namque, quod minor debeatur pena furanti in loco communi, quam in privato, quia ipsa ordinatur proportionaliter in peccatum. Sed ille, qui furatur in loco communi, minus peccat eo, qui furatur in proprio. Quod probatur, quia ubi maior datur occasio peccandi, ibi debet minor pena infigi, ut apparuiut quarto precedenti [*particule*]."

**84** *Expositio*, 29.13 [29.14]: "Nam si adolescens puniatur, minori debet puniri pena quam senex." See also *Problem* 28.4.

**85** *Expositio*, 29.13 [29.14]: ". . . ad probationem respondendum, quod, licet in loco communi prima fronte maior videatur esse causa et inclinatio ad peccandum, non tamen secundum veritatem. Considerando ulterius penam mortis, quam lator legis instituit, si quis in hoc deprehendatur ingenians se in modum nature ad excessum uniuscuiusque."

**86** *Expositio*, 29.13 [29.14]: "Et confirmatur, quia dicunt deum non fuisse misertum peccati Luciferi, ut eius, quod hominis, quia ille nullam causam extra se habuit incitativam ad peccandum. Vir autem ipsam habuit et mulier." Apparently, Lucifer did not know the penalty.

the distinction between willed and unwilled action as either primary or especially useful for the analysis of action. His lack of interest in this distinction is clear where the text of the *Problemata* appears to beg such an analysis. To *Problem* 30.11, "Why does man, especially, think one thing and do another?" Pietro gives only a cursory commentary.[87] This issue is central to the discussion in Book 7 of the *Ethica Nicomachea* about the failure displayed by incontinent people, who seem to somehow act against the judgement of their will and reason. Yet, Pietro neither cites Aristotle's treatment nor does he reveal familiarity with the considerable medieval discussion.[88] Instead, he sidesteps the difficult by observing that, in any event, most people do not have an intellect capable of right reason; to the contrary, people are impelled by their passions in both their reasoning and their actions.[89] Complex psychological discussions about the failure of the will or intellect have no place here. In fact, throughout the *Expositio*, the texts from the *Ethica Nicomachea* and other ethical works are limited to providing structure and defining terms. Pietro never raises points which might challenge his own physiological perspective; he is a very selective reader of his sources. Points of contest and doubt have a physiological basis and originate almost entirely from medical works.[90]

This understanding of moral philosophy is a deliberate and comprehensive revision of the domain of ethical inquiry, restoring the animality and sensuality lacking in most scholastic analyses of action through his emphasis on the sensitive soul. The effect of this perspective is to blur the distinction between the natural and moral orders; this approach to human action that was one of the primary concerns about medical philosophy for some theologians: namely that the actions of the soul could be naturally determined. If human agents are incapable of acting outside of the natural order, actions would lose their morality.[91] The opposite seems to be the case in Pietro's commentary. Instead, even the most mundane aspects of human existence have the potential to become moralized: diet, regimen, and indeed any activity that affects the body's qualitative balance become critical for virtuous behavior.

---

**87** *Problemata*, 30.11: "Propter quid aliud intelligit et facit homo maxime?"

**88** For medieval discussion of this issue, see Saarinen 1994.

**89** *Expositio*, 30.11: "[H]omo est animal deceptum et maxime, cum fuerit depravatum et a ratione semotum . . . Dubitatur quia dicit hominem ut plurimum 'vivere secundum intellectum' . . . Vel potest dici quae 'vivere secundum intellectum' quia utatur actibus illius sive sit rectus sive depravatus, vel quomodolibet aliter dummodo utatur eo. Et ita plures vivunt secundum intellectum. Excipiuntur enim pueri et amentes omnifariam ut occurantur illud."

**90** There are good examples of this in *Problems* 27.2, 27.9, and 28.7. In these commentaries, he challenges information in the Aristotelian text with lengthy digressions taken from medical doctrine.

**91** See: Jacquart 2001: 40–41, for some of the condemnations from 1277 which address these concerns.

# Epilogue

In differentia 9 of the *Conciliator,* Pietro d'Abano paints a bleak picture of the current state of human nature.[1] Like a grouchy old man, he rants that people are not what they once were. Although he acknowledges that there might be arguments to the contrary, he argues that since antiquity, human nature has been corrupted. Once upon a time, we were like giants: stronger, larger, more beautiful, longer-living, and farther-seeing. People were temperate: "In ancient times, men were content with morsels and chestnuts, since men used what due (*debita*), in the administration of their diets and food." Natures were strong and humors were more balanced and resistant to flux (*evulsioni*). People copulated when they were supposed to: male and female only "conjoined in a perfect state", when men were 35 and women 18, which resulted in sturdier offspring. People could tolerate stronger medicines, and properly moderated their regimen of the six non-naturals. They even had more teeth back then! (32 instead of 28).

But nowadays, "human nature has deteriorated". There is too much "cupidity, greed, gluttony, and lust, and the proper regimen of health is not being followed". There are beggars everywhere. It used to be that eunuchs and young boys were never afflicted with arthritis (according to Hippocrates' *Aphorisms* 6.28 and 30), but already by the time of Galen this had changed because of all this soft living. Our manner of procreation is corrupting us: people copulate before the due time or in the wrong manner. With sex outside of marriage, "the offspring will be ruled from whence it came: it is depraved – thus nature makes the offspring vile." Not only that, but many have disordered dispositions, filled with food and drink, and "drunk people generate only weakly". People have inordinate regimens, eating too many elegant foods, drinks and delicacies. It can only be concluded, Pietro finishes, that our nature is now indolent (*imbellicitatem*) compared to what it once was.

The causes of all this physical and moral degradation are numerous, some of them more in our control than others. Obviously, there are the celestial forces, from the motions of the eighth sphere to all the planets. "Human nature", he says, has been "perverted because of the influence of the ruling planet on complexion". Conjunctions of Saturn and Jupiter have been particularly pernicious. Closer to home, sometimes the elements have been corrupted, with one element overwhelming the others with floods (as happened in the time of Noah), vapors, or fires. Other times, it is our own way of life that causes our decline. All of this corrupts our complexions.

However, he ends this opening section of the *Conciliator* on an optimistic note. The whole ambition of Pietro's medical anthropology is the reversal of all this deca-

1 *Conciliator*, diff. 9 (ff. 14r-15v): "Utrum natura humana sit debilitata ab eo quod antiquitus, necne?" (Whether human nature has deteriorated from what it was in antiquity?). All examples below come from *Propter* 3, where Pietro provides his own answer, unless otherwise noted.

https://doi.org/10.1515/9783110612639-008

dence. The goal of this science "is the perfection of the human soul, preparing it in effect for future happiness".[2] We can understand and harness the influences of the stars and we can correct our environment and manner of living. The causes of our current degradation are natural and, for Pietro dAbano, "where nature is the cause, medicine is the answer", as Joan Cadden aptly puts it.[3] This book has aimed to show many of the ways that body and soul are connected in Pietro's medical anthropology, laying the foundation for practical medical solutions to all our ills, mental and physical.

**2** Diff. 1, prop. 3.
**3** Cadden 2013: 69.

# Bibliography

## Sources

### Manuscripts

**Pietro d'Abano, *Expositio Problematum Aristotelis***
Cesena, Biglioteca Malatesiana, XXIV.Dext.2
Nürnberg, Staatsbibliothek, Cent.III.38
Paris, Bibliotheque nationale de France, latin 6541

**Pietro d'Abano, *Liber Compilationis Physionomie***
Vatican, *Barberinianus latinus* 341

## Editions

**Adelard of Bath, *De eodem et diverso, Questiones naturales, and De avibus tractatus***, ed. and transl. Charles Burnett with Italo Ronca, Pedro Mantas Espana, and Baudouin van den Abeele, in *Conversations with his nephew: On the same and the different, Questions on Natural Science, and On Birds.* (Cambridge Medieval Classics 9). Cambridge: Cambridge University Press, 1998.

**Albertus Magnus, *De animalibus***, ed. Hermann Stadler (Beiträge zur Geschichte der Philosophie des Mittelalters 15 and 16). Münster: Aschendorff, 1916 and 1920.
Engl. transl. Kennth Kitchell Jr., and Irven Michael Resnick, *On Animals: A Medieval Summa Zoologica*, 2 vols. Columbus: Ohio State University Press, 2018.

**Albertus Magnus, *Quaestiones super de animalibus***, ed. Ephrem Filthaut in *Opera omnia*, vol. XII. Münster: Aschendorff, 1955.
Engl. transl. Irven Michael Resnick, and Kennth Kitchell Jr., *Questions Concerning Aristotle's On Animals.* Washington, D.C.: Catholic University of America Press, 2008.

**Al-Ghazzali, *Metaphysics***, ed. Latin medieval transl. Joseph T. Muckle. Toronto: St. Michael's College, 1933.

**Aristotle, *De Anima***, ed. and transl. Walter Stanley Hett (Loeb Classical Library 288). Cambridge, MA: Harvard University Press, 1936.

**Aristotle, *De generatione animalium***, ed. and transl. Arthur Leslie Peck (Loeb Classical Library 366). Cambridge, MA: Harvard University Press, 1942.
Latin medieval translations:
ed. Joan Hendrik Drossart Lulofs, *Guillelmi de Morbeka, De generatione animalium* (Aristoteles Latinus XVII, 2.V, 2 vols.). Bruges-Paris: De Brouwer, 1966.
ed. Aafke M.I. van Oppenraaij, *Aristotle De animalibus, Michael Scot's Arabic-Latin translation.* Part three, books XV–XIX, *Generation of animals*, with a Greek index to *De generatione animalium* by Hendrik Joan Drossart Lulofs (Aristoteles Semitico-Latinus 5.3). Leiden: Brill, 1992.

**Aristotle, *De motu animalium***, ed. and transl. Martha Nussbaum. Princeton: Princeton University Press, 1978.

**Aristotle, *De partibus animalium***, ed. and transl. Arthur Leslie Peck (Loeb Classical Library 323). Cambridge, MA: Harvard University Press, 1961.

https://doi.org/10.1515/9783110612639-009

Latin medieval translation:

ed. Aafke M.I. van Oppenraaij, *Aristotle, De animalibus. Michael Scot's Arabic-Latin Translation*. Part two, books XI–XIV, *Parts of Animals*. (Aristoteles Semitico-Latinus 5.2). Leiden: Brill, 1998.

**Aristotle, *Ethica Nicomachea***, ed. and transl. Harris Rackham (Loeb Classical Library 73). Cambridge, MA: Harvard University Press, 1926.

Latin medieval translation:

ed. René Antoine Gauthier, *Translatio Roberti Grosseteste Lincolniensis sive "Liber ethicorum." Recensio pura*. (Aristoteles Latinus 26/3). Leiden: Brill, 1972.

**Aristotle, *Historia animalium***, ed. and transl. Arthur Leslie Peck (vols 1 and 2), ed. and transl. David Mowbray Balme (vol. 3) (Loeb Classical Library 437–439). London and Cambridge, MA: Harvard University Press, 1965, 1970, and 1991.

Latin medieval translations:

ed. Aafke M.I. van Oppenraay, *Aristotle, De animalibus. Michael Scot's Arabic-Latin Translation*. Part one A, *Books I–III: History of Animals*. With a Description of the Base Manuscript by Erik Kwakkel. (Aristoteles Semitico-Latinus 5.1a). Leiden: Brill, 2020.

eds Pieter Buellens and Fernand Bossier, *De historia animalium. Translatio Guillelmi de Morbeka* (Aristoteles Latinus XVII 2.I, 2 vol.). Turnhout: Brepols, 2000 and 2021.

**Aristotle, *Meteorologica***, ed. and transl. Henry Desmond Pritchard Lee (Loeb Classical Library 397). Cambridge, MA: Cambridge University Press, 1952.

**Aristotle, *Parva naturalia***, ed. and transl. Walter Stanley Hett (Loeb Classical Library 288). Cambridge, MA: Harvard University Press, 1936.

**Aristotle, Physica**, transl. Philip H. Wicksteed and Francis Macdonald Cornford (Loeb Classical Library 228). Cambridge, MA: Harvard University Press, 1957.

**(Aristotle), *Physiognomy***, ed. and transl. Simon Swain, in *Seeing the Face, Seeing the Soul: Polemon's Physiognomy from Classical Antiquity to Medieval Islam*. Oxford: Oxford University Press, 2007, pp. 637–661.

Latin medieval translation:

ed. Bartholomew of Messina, Richard Foerster, in *Scriptores physiognomonici graeci et latini*. Leipzig: Teubner, 1893, vol. 1, pp. 2–91.

**(Aristotle), *Problemata***, ed. and transl. Walter Stanley Hett, 2 vol (Loeb Classical Library 316–317). Cambridge, MA: Harvard University Press, 1936–1937. (ed. and French transl. Pierre Louis, 3 vol. Paris: Belle Lettres, 1991.)

**Averroes (ibn Rushd), *Commentarium magnum in Aristotelis De anima libros***, ed. Latin medieval transl. Frederick Stuart Crawford. Cambridge, MA: Medieval Academy of America, 1953.

**Avicenna (ibn Sina), *Liber Canonis***. Venice: Paganium de Paganinis, 1507. Reprint Hildesheim: G. Olms, 1964.

**Avicenna (ibn Sina), *Liber de anima seu sextus de naturalibus***, ed. Simone Van Riet (Avicenna Latinus). Louvain: Peeters and Leiden: Brill, 1968 (I–III) and 1972 (IV–V).

**Bernardus Silvestris, *Cosmographia***, ed. and transl. Winthrop Wetherbee in *Poetic Works* (Dumbarton Oaks Medieval Library 38). Cambridge, MA: Harvard University Press, 2015.

***Chartularium Universitatis Parisiensis***, ed. Heinrich Denifle and Emile Chatelain. Paris: Delallain, 1889–1897.

**Constantine the African, *Pantegni***, in *Opera omnia Ysaac*, vol. 2, ff. 1r–144r. Lyon: no name, 1515.

**Francesco Petrarca, *Invectives***, ed. and Engl. transl. David Marsh (I Tatti Renaissance Library 2). Cambridge, MA: Harvard University Press, 2003.

**Galen, *Tegni***, ed. and transl. Ian Johnston, *The Art of Medicine* (Loeb Classical Library 523). Cambridge, MA: Harvard University Press, 2016.

**Galen, *De complexionibus***, ed. Latin transl. Richard Durling, *Burgundio of Pisa's Translation of Galen's Peri Kraseon (De Complexionibus)* (Ars Medica 6.1). Berlin and New York: De Gruyter, 1976.

Engl. transl. Peter Singer and Philip J. van Der Eijk, *Galen: Works on Human Nature, vol. 1. Mixtures.* Cambridge: Cambridge University Press, 2018.

**Galen, *Selected Works***, ed. and transl. Peter N. Singer. Oxford: Oxford University Press, 1997.

**Haly Abbas** (Ali ibn al-'Abbas al-Magusi). See: Constantine, *Pantegni*.

**Hippocrates, *Aphorisms***, ed. and transl. William Henry Stuart Jones (Loeb Classical Library 150). Cambridge, MA: Harvard University Press, 1931.

**Hippocrates, *Prognostic***, ed. and transl. Paul Potter (Loeb Classical Library 148). Cambridge, MA: Harvard University Press, 2023.

**Hugh of St. Victor, *Didascalicon***, transl. Jerome Taylor, *The Didascalicon of Hugh of St. Victor: A Medieval Guide to the Arts*. New York: Columbia University Press, 1961.

**Isidore of Seville, *Etymologies***, ed. Wallace Lindsey, *Isidori Hispalensis Episcopi Etymologiarum Sive Originum Libri XX*. Oxford: Oxford University Press, 1911.

Engl. transl. Stephen Barney, W.J. Lewis, J.A. Beach, and Oliver Berghof, *The Etymologies of Isidore of Seville*. Cambridge: Cambridge University Press, 2006.

**Justinian, *Institutes***, ed. Paul Krüger, in Paul Krüger and Theodor Mommsen (eds), *Corpus Iuris Civilis*. Berlin: Weidmannos, 1872.

**Johannitius (Hunayn ibn Ishāq), *Isagoge***, ed. Gregor von Maurach, *Sudhoffs Archiv* 62 (1978), pp. 148–74. Engl. transl. in Faith Wallis, *Medieval Medicine: A Reader*. Toronto: University of Toronto Press, 2010, pp. 139–159.

**John of St. Amand, *Concordantiae***, ed. Julius Leopold Pagel, *Die Concordanciae des Johannes de Sancto Amando*. Berlin: Georg Reimer, 1894. Reprint, Berlin and Boston: De Gruyter, 2019.

**Nemesius, *De natura hominis***, ed. Moreno Morani (Bibliotheca Graeca et Romana Teubneriana). Leipzig: Teubner, 1987.

Latin medieval translation:

eds Gerard Verbeke et Jose Rafael Moncho, *Némésius d'Émèse,* De natura hominis, *traduction de Burgundio de Pise* (Corpus Latinum Commentariorum in Aristotelem Graecorum, suppl 1).. Leiden: Brill, 1975.

Engl. transl. with an introduction and notes, Robert Sharples and Philip van der Eijk, *On the Nature of Man*. Liverpool: Liverpool University Press, 2008.

**Peter of Spain, *Questiones super Libro De Animalibus Aristotelis***, ed. Francesca Navarro Sanchez, *Peter of Spain, Questiones super Libro 'De Animalibus' Aristotelis. Critical Edition with Introduction* (Medicine in the Medieval Mediterranean 5). New York: Routledge, 2015.

**Peter of Spain, *Thesaurus Pauperum***, ed. and Spanish transl. Maria Helena da Rocha Pereira. Coimbra: University of Coimbra Press, 1973.

**Pietro d'Abano, *Conciliator***. Venice: Iuntas, 1565. Reprint Padua: Antenor, 1985.

*Differentia 135 and 156*, ed. Kurt Boughan in *Beyond Diet, Drugs, and Surgery: Italian Scholastic Medical Theorists on the Animal Soul, 1270–1400*. 2 vol. PhD diss., University of Iowa, 2006, vol. 2, pp. 281–326.

*Differentia 156*, ed. Béatrice Delaurenti, "Pietro d'Abano et les incantations. Présentation, édition, et traduction de la differentia 156 du Conciliator," in Jean-Patrice Boudet, Franck Collard, and Nicolas Weill-Parot (eds), *Médecine, astrologie, et magie: autour de Pietro d'Abano* (Micrologus Library 50). Florence: SISMEL Edizioni del Galluzzo, 2012, pp. 39–105.

**Pietro d'Abano, *Expositio Problematum Aristotelis***. Mantua: P. de Butzbach, 1475.

*Prologue*, ed. Pieter De Leemans, "*Ego, Petrus Paduanus, philosophie minimus alumpnorum*. Pietro d'Abano's Preface to his *Expositio Problematum Aristotelis*," in Pieter de Leemans and Maarten Hoenen (eds), *Between Text and Tradition. Pietro d'Abano and the Reception of Pseudo-Aristotle's* Problemata physica *in the Middle Ages*. Leuven: Leuven University Press, 2016, pp. 21–52.

*Book 4*, ed. Gijs Coucke, "Philosophy between Text and Tradition The Reception of Aristotle's *Problemata* in the Middle Ages: Peter of Abano's *Expositio Problematum* Volume II. Edition of

Peter of Abano, Exp., IV, with Synthesis, Annotations and Essays." PhD Thesis, Catholic University of Leuven, 2008.

*Book 7*, ed. Kurt Boughan in *Beyond Diet, Drugs, and Surgery: Italian Scholastic Medical Theorists on the Animal Soul, 1270–1400*. 2 vol. PhD diss., University of Iowa, 2006, vol. 2, pp. 327–379.

*Book 19*, ed. Christian Meyer, *Pietro d'Abano,* Expositio problematum *(XIX). Édition, introduction et notes critiques et explicatives*. Leuven: Leuven University Press, 2022.

*Book 32*, ed. Lise Devriese, "Aristotle on the Ears. A Study and Critical Edition of Bartholomew of Messina's Translation and Peter of Abano's Commentary on Aristotle's *Problemata physica*." MA Thesis, Catholic University of Leuven, 2013.

**Pietro d'Abano, *Liber compilationis physionomie***. Padua: no name, 1474.

**Pietro d'Abano, *Lucidator***, ed. Graziella Federici Vescovini, *Il "Lucidator dubitabilium astronomiae" di Pietro d'Abano: opere scientifiche inedited*. Padua: Programma e 1+1, 1988.

**Plato, *Timaeus a Calcidio translatus commentarioque instructus***, ed. Jan Hendrik Waszink. Leiden and London: Brill and the Warburg Institute, 1962.

**Polemon, *Physiognomy***, ed. Latin transl. George Hoffman in *Scriptores physiognomonici graeci et latini* (Bibliotheca Graeca et Latina Teubneriana). Leipzig: Teubner, 1893, vol. 1, pp. 93–294.

**Rhazes (al-Rāzī), *physiognomy***, Latin transl. Gerard of Cremona, ed. Richard Foerster in *Scriptores physiognomonici graeci et latini* (Bibliotheca Graeca et Latina Teubneriana), Leipzig: Teubner, 1893, vol. 2, pp. 161–179.

**Roger Bacon, *De erroribus medicorum***, ed. Edward Withington, in *Opera hactanus inedita Rogeri Baconi*, fasc. 9: *De retardatione accidentium senectutis cum aliis opusculis de rebus medicinalibus*. Oxford: Oxford University Press, 1928.

**Roger Bacon, *Secretum secretorum cum glossis et notulis***, ed. Robert Steele, in *Opera hactenus inedita Rogeri Bacon*, fasc. 5. Oxford: Oxford University Press, 1920.

***Scriptores physiognomonici graeci et latini***, ed. Richard Foerster, 2 vols. Leipzig: Teubner, 1893.

**Taddeo Alderotti, *Expositiones in arduum aphorismorum Ippocrates volumen***. Venice: Luca Antonio Giunta, 1527.

**Thomas Aquinas, *Summa Theologiae***, Engl. transl. Alfred J. Freddoso: https://www3.nd.edu/~afreddos/summa-translation/TOC.htm (last access 07.08.2024).

**Thomas Aquinas, *Quaestiones disputatae de anima***, ed. Bernardo Carlos Bazán, in *Opera Omnia* 24.1. Roma: Commissio Leonina; Paris: Cerf, 1996.

**Urso of Salerno, *Aphorisms***, ed. Rudolf Creutz, "Die medizinische-naturphilosophischen Aphorismen und Kommentare des Magister Urso Salernitanus," *Quellen und Studien zur Geschichte der Naturwissenschaften und der Medizin* 5–1 (1936), pp. 1–192.

**Urso of Salerno, *De commixtionibus elementorum libellis***, ed. Wolfgang Stürner. Stuttgart: Klett, 1976.

**William of Conches, *De philosophia mundi***, ed. Gregor Maurach with Heidemarie Telle. Pretoria: University of South Africa Press, 1980.

**William of Conches, *Dragmaticon***, ed. Italo Ronca, in *Guillelmi de Conchis Dragmaticon philosophiae*. Opera omnia 1. (Corpus Christianorum Continuatio Mediaevalis 152). Turnhout: Brepols, 1997.

Engl. transl. Matthew Curr and Italo Ronca, *Dialogue on Natural Philosophy* (*Dragmaticon Philosophiae*) William of Conches. South Bend: Notre Dame University Press, 1997.

## Literature

**Aertsen, Emery, and Speer 2001**: Johannes Adrianus Aertsen, Kent Emery, $1 Andreas Speer (eds), *Nach der Verurteilung von 1277. Philosophie und Theologie an der Universität von Paris im letzten Viertel des 13. Jahrhunderts. Studien und Texte*. Berlin and New York: De Gruyter, 2001.

**Agrimi and Crisciani 1978**: Jole Agrimi and Chiara Crisciani, *Medicina del corpo e medicina dell'anima: Note sul sapere del medico fino all'inizio del sec. XIII*. Milan: Episteme Editrice, 1978.

**Agrimi and Crisciani 1988**: Jole Agrimi and Chiara Crisciani, *"Edocere medicos." Medicina scolastica nei secoli XIII–XV*. Naples: Guerini, 1988.

**Agrimi and Crisciani 1994**: Jole Agrimi and Chiara Crisciani, *Les "consilia" médicaux*. Turnhout: Brepols, 1994.

**Agrimi and Crisciani 1998**: Jole Agrimi and Chiara Crisciani, "Charity and Aid in Medieval Christian Civilization," in Mirko Grmek (ed.), *Western Medical Thought from Antiquity to the Middle Ages*. Translated by Antony Shugaar. Cambridge, MA and London: Harvard University Press, 1998, pp.170–196.

**Alessio 1976**: Franco Alessio, "Filosofia e scienza in Pietro d'Abano," in Franco Alessio. Girolamo Arnaldi, and Manlio Stocchi (eds), *Storia della cultura veneta*, Vol. 2. Vicenza: Neri Pozza, 1976, pp. 171–200.

**Amerini 2009**: Fabrizio Amerini, *Tommaso d'Aquino. Origine e fine della vita humana* (Philosophica, 58). Pisa: ETS, 2009.

**Amundsen 1996**: Darrel Amundsen, *Medicine, Society, and Faith in the Ancient and Medieval Worlds*. Baltimore and London: Johns Hopkins University Press, 1996.

**Baldwin 1970**: John Baldwin. *Masters, Princes and Merchants: The Social Views of Peter the Chanter and his Circle*. Princeton: Princeton University Press, 1970.

**Beccaria 1956**: Augusto Beccaria, *I codici di medicina del period presalernitano (secoli IX, X, e XI)*. Rome: Edizioni di storia e letteratura, 1956.

**Bianchi 1987**: Luca Bianchi, "La felicità intelletuale come professione nella Parigi del Duecento," *Rivista di filosofia* 78 (1987), pp. 181–200.

**Bianchi 1998**: Luca Bianchi, "1277: A Turning Point in Medieval Philosophy?," in Jan A. Aertsen and Andreas Speer (eds), *Was ist Philosophie im Mittelalter?* Berlin and New York: De Gruyter, pp. 90–110.

**Bianchi 2003**: Luca Bianchi, "New Perspectives on the Condemnation of 1277 and its Aftermath," *Recherches de théologie e philosophie médiévale* 70 (2003), pp. 206–229.

**Birkenmajer 1970**: Aleksander Birkenmajer, "Le rôle joué par les médecins et les naturalistes dans la réception d'Aristote au XIe et XIIIe siècles," *Etudes d'histoire des sciences et de la philosophie du moyen âge. Studia Copernicana* 1 (1970) [1930], pp. 73–87.

**Black 1999**: Deborah Black, "The Nature of the Intellect," in Robert Pasnau and Christina Van Dyke (eds), *The Cambridge History of Medieval Philosophy*. Cambridge: Cambridge University Press, 1999, pp. 305–319.

**Blair 1999**: Ann Blair, "The Problems as a Natural Philosophical Genre," in Anthony Grafton and Nancy Siraisi (eds), *Natural Particulars: Nature and the Disciplines in Renaissance Europe*. Cambridge, MA: MIT Press, 1999, pp.171–204.

**Bono 1984**: James J. Bono, "Medical Spirits and the Medieval Language of Life," *Traditio* 40 (1984), pp. 91–130.

**Bottin 2013**: Francesco Bottin, "Peter of Abano: The Scientific Method of a 'Diligens Indagator'," *Medioevo* 38 (2013), pp. 317–360.

**Boudet, et al. 2013**: Jean-Patrice Boudet, Franck Collard, and Nicolas Weill-Parot (eds), *Médicine, astrologie et magie entre Moyen Âge et Renaissance: autour de Pietro d'Abano* (Micrologus Library 50). Florence: SISMEL Edizioni del Galluzzo, 2013.

**Boughan 2006**: Kurt Boughan. *Beyond Diet, Drugs, and Surgery: Italian Scholastic Medical Theorists on the Animal Soul, 1270–1400*. 2 vol. PhD diss., University of Iowa, 2006.

**Bouras-Vallianatos and Zipser 2019**: Petros Bouras-Vallianatos and Barbara Zipser (eds), *Brill's Companion to the Reception of Galen* (Brill's Companions to Classical Reception 17). Leiden: Brill, 2019.

**Boys-Stones 2007**: George Boys-Stones, "Physiognomy and Ancient Psychological Theory," in Simon Swain (ed.), *Seeing the Face, Seeing the Soul: Polemon's Physiognomy from Classical Antiquity to Medieval Islam*. Oxford: Oxford University Press, 2007, pp. 19–124.

**Buellens 2007**: Pieter Buellens, “Like a Book Written by God’s Finger: Animals Showing the Path toward God,” in Brigitte Resl (ed.), *A Cultural History of Animals*, Vol. II: *The Medieval Age*. Oxford: Berg, 2007, pp. 127–151.

**Buellens 2016**: Pieter Buellens, “Burgundio of Pisa,” in Thomas Glick, Steven J. Livesey, and Faith Wallis (eds), *Medieval Science, Technology, and Medicine*. New York, NY: Routledge, 2016, pp.104–105.

**Bullough 1956**: Vern L. Bullough, “The Development of the Medical University at Montpellier to the End of the Fourteenth Century,” *Bulletin of the History of Medicine* 30 (1956), pp. 508–523.

**Bullough 1957**: Vern L. Bullough, “The Medieval Medical University in Paris,” *Bulletin of the History of Medicine* 31 (1957), pp. 197–211.

**Bürgel 1967**: J. Cristoph Bürgel, *Averroes “Contra Galenum,”* Göttingen: Vandenhoeck and Ruprecht, 1967.

**Burnett 1984**: Charles Burnett, “The Contents and Affiliation of the Scientific Manuscripts Written at, or Brought to, Chartres in the Time of John of Salisbury,” in Michael Wilks (ed.), *The World of John of Salisbury*. Oxford: Blackwell, 1984, pp. 127–160.

**Burnett 1988**: Charles Burnett, “Scientific Speculations,” in Peter Dronke (ed.), *A History of Twelfth-Century Western Philosophy*. Cambridge: Cambridge University Press, 1988, pp. 151–176.

**Burnett 1994**: Charles Burnett, “The Chapter on the Spirits in the *Pantegni* of Constantine the African,” in Charles Burnett and Danielle Jacquart (eds), *Constantine the African and ‘Alī al-‘Abbās al-Mağūsī: The* Pantegni *and Related Texts* (Studies in Ancient Medicine 10). Leiden: Brill, 1994, pp. 99–120.

**Burnett 1997**: Charles Burnett, “Hearing and Music in Book XI of Pietro d’Abano’s *Expositio Problematum Aristotelis*,” in Nancy van Deusen (ed.), *Tradition and Ecstasy: The Agony of the Fourteenth Century*, Ottawa: Institute of Medieval Music, 1997, pp. 153–190.

**Burnett 2007**: Charles Burnett, “*Verba Ypocratis preponderanda omnium generum metallis*. Hippocrates on the Nature of Man in Salerno and Montecassino, with an Edition of the Chapter on the Elements in the *Pantegni*,” in Danielle Jacquart and Agostino Paravicini Bagliani (eds), *La Scuola Medica Salernitana. Gli autori e i testi* (Edizione Nazionale “La Scuola Medica Salernitana” 1). Florence: SISMEL Edizioni del Galluzzo, 2007, pp. 59–92.

**Burnett 2013a**: Charles Burnett, “Translation and Transmission of Greek and Islamic Science to Latin Christendom,” in David Lindberg and Michael Shank (eds), *The Cambridge History of Science*, vol. 2: *Medieval Science*. New York, NY: Cambridge University Press, 2013, pp. 341–364.

**Burnett 2013b**: Charles Burnett, “The Twelfth-Century Renaissance,” in David Lindberg and Michael Shank (eds), *The Cambridge History of Science*, vol. 2: *Medieval Science*. New York, NY: Cambridge University Press, 2013, pp. 365–384.

**Burnett 2016a**: Charles Burnett, “Gerard of Cremona,” in Thomas Glick, Steven J. Livesey, and Faith Wallis (eds), *Medieval Science, Technology, and Medicine*. New York, NY: Routledge, 2016, pp. 191–192.

**Burnett 2016b**: Charles Burnett, “Michael Scot,” in Thomas Glick, Steven J. Livesey, and Faith Wallis (eds), *Medieval Science, Technology, and Medicine*. New York, NY: Routledge, 2016, pp. 344–345.

**Burnett and Jacquart 1994**: Charles Burnett and Danielle Jacquart (eds), *Constantine the African and ‘Alī al-‘Abbās al-Mağūsī: The* Pantegni *and Related Texts* (Studies in Ancient Medicine 10). Leiden: Brill, 1994.

**Bylebyl 1971**: Jerome J. Bylebyl, “Galen on the Non-Natural Causes of Variation in the Pulse,” *Bulletin of the History of Medicine* 45 (1971), pp. 482–485.

**Bylebyl 1990**: Jerome J. Bylebyl, “The Medical Meaning of Physica,” *Osiris*, 6 (1990), pp. 16–41.

**Cadden 1995**: Joan Cadden, “Science and Rhetoric in the Middle Ages: The Natural Philosophy of William of Conches,” *Journal of the History of Ideas* 56 (1995), pp. 1–24.

**Cadden 2001**: Joan Cadden, “’Nothing natural is shameful’: Vestiges of a Debate About Sex and Science in a Group of Late-Medieval manuscripts,” *Speculum* 76 (2001), pp. 66–89.

**Cadden 2013**: Joan Cadden, *Nothing Natural is Shameful: Sodomy and Science in Late-Medieval Europe*. Philadelphia, PA: University of Pennsylvania Press, 2013.

**Cadden 2016**: Joan Cadden, "Sex and Sensibilities in the Medieval *Problemata* Tradition: Pietro d'Abano and His Readers," in Pieter De Leemenas and Maarten Hoenen (eds), *Between Text and Tradition: Pietro d'Abano and the Reception of Pseudo-Aristotle's* Problemata Physica *in the Middle Ages*. Leuven: Leuven University Press, 2016, pp. 53–80.

**Carbone 2010**: Andrea Carbone, "Anomolies de l'intelligence, intelligence de l'anomalie," in Pierre-Marie Morel and Cristophe Grellard (eds), *Les Parva naturalia d'Aristotle: Fortune antique et médiévale*. Paris: Éditions de la Sorbonne, 2010, pp. 11–45.

**Celano 1986**: Anthony J. Celano, "The 'finis hominis' in the Thirteenth-Century Commentaries on Aristotle's *Nichomachean Ethics*," *Archive d'histoire doctrinale et littéraire du moyen âge* 53 (1986), pp. 23–53.

**Celano 1990**: Anthony J. Celano, "Act of the Intellect or Act of the Will: The Critical Reception of Aristotle's Ideal of Human Perfection in the 13th and Early 14th Centuries," *Archive d'histoire doctrinale et littéraire du moyen âge* 57 (1990), pp. 93–119.

**Chandelier 2013**: Joël Chandelier, "Pietro d'Abano et les médecins: réception et réputation du *Conciliator* en Italie dans les premières années du XIV e siècle," in Jean-Patrice Boudet, Franck Collard, and Nicolas Weill-Parot (eds), *Médicine, astrology et magie entre Moyen Âge et Renaissance: autour de Pietro d'Abano* (Micrologus Library 50). Florence: SISMEL Edizioni del Galluzzo, 2013, pp. 183–201.

**Chandelier and Robert 2013**: Joël Chandelier and Aurélien Robert, "Nature humaine et complexion du corps chez les médicins Italiens de la fin du moyen âge," *Revue de synthèse* 134 (2013), pp. 473–510.

**Compagni 2009**: Vittoria Perrone Compagni, "La differenza 156 del Conciliator: una rilettura," *Annali del Dipartimento di Filosofia (Nuova Serie)*, 15 (2009), pp. 65–107.

**Cooper 2011**: Glen M. Cooper, *Galen*, De diebus decretoriis, *from Greek into Arabic: A Critical Edition, with Translation and Commentary*, of Hunayn ibn Ishaq, Kitab ayyam al-buhran (Medicine in the Medieval Mediterranean 6). Farnham and Burlington, VT: Ashgate, 2011.

**Coucke 2016**: Gijs Coucke, "From Abstinence to Promiscuity: Men, Beasts, and Eunuchs in the *Expositio Problematum* of Pietro d'Abano," in Pieter De Leemenas and Maarten Hoenen (eds), *Between Text and Tradition: Pietro d'Abano and the Reception of Pseudo-Aristotle's* Problemata Physica *in the Middle Ages*. Leuven: Leuven University Press, 2016.

**Coucke and Swaenepoel 2009**: Gijs Coucke and Tine Swaenepoel. "The Relation between Bartholomew's Translation of the *Problemata* and Peter of Abano's *Expositio Problematum*," in Pieter De Leemans and Carlos Steel (eds), *The Aristoteles Latinus Past, Present and Future*. Brussels: Koninklijke Vlaamse Academie van België voor Wetenschappen en Kunsten, 2009.

**Crisciani 1983**: Chiara Crisciani, "Valeurs éthiques et savoir médical entre le XIIe et le XIVe siècle," *History and Philosophy of the Life Sciences* 5 (1983), pp. 33–52.

**Dales 1995**: Richard Dales, *The Problem of the Rational Soul in the Thirteenth Century*. Leiden: Brill, 1995.

**d'Alverny 1966**: Marie-Thérése d'Alverny,"Pietro d'Abano et les 'naturalistes' a l'epoque de Dante," in Vittoria Branco and Giorgio Padoan (eds), *Dante e la cultura veneta*. Florence : Olschki, 1966, pp. 207–219.

**d'Alverny 1982**: Marie-Thérése d'Alverny, "Translations and Translators," in Robert Benson and Giles Constable (eds), *Renaissance and Renewal in the Twelfth Century*. Cambridge, MA: Harvard University Press, 1982, pp. 421–462.

**d'Alverny 1985**: Marie-Thérése d'Alverny, "Pietro d'Abano traducteur de Galien," *Medioevo* 11 (1985), pp. 19–64.

**Daston and Park 1998**: Lorraine Datson and Katharine Park, *Wonders and the Order of Nature, 1150–1750*. New York: Zone Books, 1998.

**De Asúa 1999**: Miguel De Asúa, "Medicine and Philosophy in Peter of Spain's Commentary on *De animlaibus*," in Carlos Steel, Guy Guldentops, and Pieter Beullens (eds), *Aristotle's Animals in the Middle Ages and the Renaissance*. Leuven: Leuven University Press, 1999, pp. 189–211.

**De Asúa 2013**: Miguel De Asúa, "War and Peace: Medicine and Natural Philosophy in Albert the Great," in Irven Resnick (ed.), *A Companion to Albert the Great: Theology, Philosophy, and the Sciences*. Leiden: Brill, 2013, pp. 269–297.

**De Boer 2013**: Sander W. De Boer. *The Science of the Soul: The Commentary Tradition on Aristotle's De Anima, c. 1260–1360*. Leuven: Leuven University Press, 2013.

**Delaurenti 2007**: Béatrice Delaurenti, *La puissance des mots.* "Virtus verborum." *Débats doctrinaux sur le pouvoir des incantations au Moyen Âge*. Paris: Cerf, 2007.

**Delaurenti 2012**: Béatrice Delaurenti, "Pietro d'Abano et les incantations. Présentation, édition, et traduction de la differentia 156 du Conciliator," in Jean-Patrice Boudet, Franck Collard, and Nicolas Weill-Parot (eds), *Médecine, astrologie, et magie: autour de Pietro d'Abano* (Micrologus Library 50). Florence: SISMEL Edizioni del Galluzzo, 2012, pp. 39–105.

**De Leemans 2006**: Pieter De Leemans, "Internal senses, intellect, and movement. Peter of Auvergne (?) on Aristotle's *De motu animalium*," in Graziella Federici Vescovini, Valeria Sorge, and Carlo Vinti, (eds), *Corpo e anima, sensi interni e intelletto dai secoli XIII–XIV ai post-cartesiani e spinoziani*. (FIDEM Textes et études du Moyen Age 30). Turnhout: Brepols, 2006, pp. 139–160.

**De Leemans 2007**: Pieter De Leemans, "Was Peter of Abano the Translator of Pseudo-Aristotle's *Problemata Physica*?" *Bulletin de Philosophie Médiévale* 49 (2007), pp. 103–118.

**De Leemans and Goyens 2006**: Pieter De Leemans and Michèle Goyens (eds), *Aristotle's* Problemata *in Different Times and Tongues*. Leuven: Leuven University Press, 2006.

**De Leemans and Hoenen 2016**: Pieter De Leemans and Maarten Hoenen (eds), *Between Text and Tradition: Pietro d'Abano and the Reception of Pseudo-Aristotle's* Problemata Physica *in the Middle Ages*. Leuven: Leuven University Press, 2016.

**De Leemans and Klemm 2007**: Pieter De Leemans and Matthew Klemm, "Animals and Anthropology in Medieval Philosophy," in Brigitte Resl (ed.), *A Cultural History of Animals*, Vol. II: *The Medieval Age*. Oxford: Berg, 2007, pp. 153–178.

**De Leemans 2016**: Pieter De Leemans, "Ego, Petrus Paduanus, philosophie minimus alumpnorum. Pietro d'Abano's Preface to his *Expositio Problematum Aristotilis*," in Pieter De Leemenas and Maarten Hoenen (eds), *Between Text and Tradition: Pietro d'Abano and the Reception of Pseudo-Aristotle's* Problemata Physica *in the Middle Ages*. Leuven: Leuven University Press, 2016, pp. 21–52.

**Demaitre 1975**: Luke Demaitre, "Theory and Practice in Medical Education at the University of Montpellier in the Thirteenth and Fourteenth Centuries," *Journal of the History of Medicine and Allied Sciences* 30 (1975), pp. 103–123.

**Demaitre 1976**: Luke Demaitre, "Nature and the Art of Medicine in the Later Middle Ages," *Medievalia* 2 (1976), pp. 23–47.

**Demaitre 2003**: Luke Demaitre, "The Art and Science of Prognostication in Early University Medicine," *Bulletin for the History of Medicine* 77 (2003), pp. 765–788.

**De Rijk 1967**: Lambertus Marie De Rijk, "Some Notes on the 12th Century Topic of the Three (Four) Human Evils and of Science, Virtue and Techniques as Their Remedies," *Vivarium* 5 (1967), pp. 8–15.

**Dod 1982**: Bernard Dod,"Aristoteles latinus." in Norman Kretzmann, Anthony Kenny, and Jan Pinborg (eds), *The Cambridge History of Later Medieval Philosophy*. Cambridge: Cambridge University Press, 1982, pp. 45–79.

**D'Ors 2001**: Angel D'Ors, "Petrus Hispanus, O.P., Auctor Summularum (II): Further Documents and Problems," *Vivarium* (2001), pp. 209–254.

**Drayton 2001**: Ralph Drayton, "Medicine and Religion in Late Medieval Culture: The Case of Astrological Talismans at the University of Montpellier," PhD diss., University of Wisconsin, Madison, 2001.

**Dronke 1988**: Peter Dronke, "Thierry of Chartres," in Peter Dronke (ed.), *A History of Twelfth-Century Western Philosophy*. Cambridge: Cambridge University Press, 1988, pp. 358–385.

**Dusenbury 2021**: David Lloyd Dusenbury, *Nemesius of Emesa on Human Nature: A Cosmopolitan Anthropology from Roman Syria*. Oxford: Oxford University Press, 2021.

**Edwards 2008**: Michael Edwards, "Digressing with Aristotle," *Early Science and Medicine* 13 (2008), pp. 127–170.

**Elford 1988**: Dorothy Elford, "William of Conches," in Peter Dronke (ed.), *A History of Twelfth-Century Western Philosophy*. Cambridge: Cambridge University Press, 1988, pp. 308–327.

**Fear and Wood 2019**: Andrew Fear and Jamie Wood (eds), *A Companion to Isidore of Seville*. Leiden: Brill, 2019.

**Ferrari 1900**: Sante Ferrari, *I tempi, la vita, le dottrine di Pietro d'Abano*. Genoa: Tipografia R. Istituto Sordomuti, 1900.

**Ferruolo 1985**: Stephen C. Ferruolo, *The Origins of the University: The Schools of Paris and their Critics, 1100–1215*. Stanford: Stanford University Press, 1985.

**Fortuna 2008**: Stephania Fortuna, "Pietro d'Abano e le Traduzioni Latine di Galeno," *Medicina nei Secoli* 20 (2008), pp. 447–463.

**French 2001**: Roger K. French, *Canonical Medicine: Gentile da Foligno and Scholasticism*. Leiden: Brill, 2001.

**French 2003**: Roger K. French, *Medicine Before Science: The Rational and Learned Doctor from the Middle Ages to the Enlightenment*. Cambridge: Cambridge University Press, 2003.

**French and Cunningham 1996**: Roger K. French and Andrew Cunningham, *Before Science: The Invention of the Friars' Natural Philosophy*. Aldershot: Scholar Press, 1996.

**García-Ballester 1988**: Luis García-Ballester, "Soul and Body. Disease of the Soul and Disease of the Body in Galen's Medical Thought," in Paola Manuli and Mario Vegetti (eds), *Le opere psicologiche di Galeno*. Naples: Bibliopolis, 1988, pp. 117–152.

**García-Ballester 1993**: Luis Garcia-Ballester, "On the Origin of the 'Six Non-Natural Things' in Galen," in Georg Harig and Jutta Kollesch (eds), *Galen und das hellenishce Erbe. Verhandlungen des IV. Internationalen Galen-symposiums veranstaltet vom Institut für Geschichte der Medizin am Bereich Medizin (Charite) der Humboldt-Universitat zu Berlin 18.–20. september 1989* (Sudhoffs Archiv Beihefte 32). Stuttgart: Franz Steiner, 1993, pp. 105–115.

**García-Ballester 1995**: Luis Garcia-Ballester, "Artifex factivus sanitatis: Health and Medical Care in Medieval Latin Galenism," in Don Bates (ed.), *Knowledge and the Scholarly Medical Traditions*. Cambridge: Cambridge University Press, 1995, pp. 127–150.

**García-Ballester 1998**: Luis Garcia-Ballester, "The New Galen: A Challenge to Latin Galenism in Thirteenth-century Montpellier," in Klaus-Dietrich Fischer, Diethard Nickel and Paul Potter (eds), *Text and Tradition: Studies in Ancient Medicine and its Transmission* (Studies in Ancient Medicine 18) Leiden: Brill, 1998, pp. 55–83.

**Gaspar and Wallis 2004**: Giles E. M. Gaspar and Faith Wallis, "Anselm and the Articella," *Traditio* 59 (2004), pp. 129–174.

**Ghersetti 2007**: Antonella Ghersetti, "The Semiotic Paradigm: Physiognomy and Medicine in Islamic Culture," in Simon Swain, (ed.), *Seeing the Face, Seeing the Soul: Polemon's Physiognomy from Classical Antiquity to Medieval Islam*. Oxford: Oxford University Press, 2007.

**Giglioni 2008**: Guido Giglioni, "Nature and Demons: Girolamo Cardano interpreter of Pietro d'Abano," in Charles Burnett, José Meirinhos, Jacqueline Hamesse (eds), *Continuities and Disruptions between the Middle Ages and the Renaissance* (Textes et Etudes du Moyen Âge 48). Turnhout: Brepols, 2008, pp. 89–112.

**Glaze 2008**: F. Eliza Glaze, "Gariopontus and the Salaernitans: Textual Traditions in the Eleventh and Twelfth Centuries," in Danielle Jacquart and Agostino Paravicini Bagliani (eds), *La 'Collectio Salernitana' di Salvatore de Renzi* (Edizione Nazionale "La Scuola Medica Salernitana" 3). Florence: SISMEL Edizioni del Galluzzo, 2008, pp. 149–190.

**Glaze 2011**: F. Eliza Glaze, "Prolegomena: Scholastic Openings to Gariopontus of Salerno's *Passionarius*," in F. Eliza Glaze and Brian K. Nance (eds), *Between Text and Patient: The Medical Enterprise in Medieval and Early Modern Europe* (Micrologus Library 39) Florence: SISMEL Edizioni del Galluzzo, 2011, pp. 57–86.

**Glaze 2019**: F. Eliza Glaze, "Introduction: Constantine the African and the *Pantegni* in Context," in Erik Kwakkel and Francis Newton, *Medicine at Monte Cassino. Constantine the African and the Oldest Manuscript of his* Pantegni. Turnhout: Brepols, 2019, pp. 1–29.

**Goddu 1985**: Andre Goddu, "The Effect of Canonical Prohibitions on the Faculty of Medicine at the University of Paris in the Middle Ages," *Medizinhistorisches Journal* 20 (1985), pp. 342–362.

**Grant 1979**: Edward Grant, "The Condemnation of 1277, God's Absolute Power, and Physical Thought in the Late Middle Ages," *Viator* 10 (1979), pp. 211–244.

**Grant 1987**: Edward Grant, "Ways to Interpret the terms "Aristotelian" and "Aristotelianism" in Medieval Natural Philosophy," *History of Science* 25 (1987), pp. 335–358.

**Green 2008**: Monica H. Green, "Rethinking the Manuscript Basis of Salvatore De Renzi's Collection Salernitana: The Corpus of Medical Writings in the 'Long' Twelfth Century," in Danielle Jacquart and Agostino Paravicini Bagliani (eds), *La 'Collection Salernitana' di Salvatore De Renzi* (Edizione Nazionale "La Scuola Medica Salernitana" 03). Florence: SISMEL Edizioni del Galluzzo, 2008, pp. 15–60.

**Green 2019**: Monica H. Green, "Gloriosissimus Galienus: Galen and Galenic Writings in the Eleventh- and Twelfth-Century Latin West," in Petros Bouras-Vallianatos and Barbara Zipser (eds), *Brill's Companion to the Reception of Galen*. Leiden: Brill, 2019, pp. 319–342.

**Guldentops 2002**: Guy Guldentops, "Henry Bate on the Potential and the Agent Intellect," *Archive d'histoire doctrinale et littéraire du moyen âge* 69 (2002), pp. 115–152.

**Hankinson 1991**: Richard. J. Hankinson, "Galen's Anatomy of the Soul," *Phronesis* 36 (1991), pp. 197–233.

**Harvey 1975**: Ruth Harvey, *The Inward Wits: Psychological Theory in the Middle Ages and the Renaissance*. London: The Warburg Institute, 1975.

**Hasse 2000**: Dag Nikolaus Hasse, *Avicenna's De anima in the Latin West*. London and Turin: Warburg Institute/Nino Aragno, 2000.

**Hasse 2001**: Dag Nikolaus Hasse, "Pietro d'Abano's 'Conciliator' and the Theory of the Soul in Paris." in Jan A. Aertsen, Kent Emery, Jr., and Andreas Speer (eds), *Nach der Verurteilung von 1277: Philosophie und Theologie an der Universität von Paris im letzten Viertel des 13. Jahrhunderts*. Berlin and New York: De Gruyter, 2001, pp. 635–653.

**Hasse 2010**: Dag Mikolaus Hasse, "The Soul's Faculties," in Robert Pasnau and Christina Van Dyke (eds), *The Cambridge History of Medieval Philosophy*. Cambridge: Cambridge University Press, 2010, pp. 305–319.

**Hirai 2011**: Hiro Hirai. *Medical Humanism and Natural Philosophy: Renaissance Debates on Matter, Life, and the Soul* (Medieval and Early Modern Philosophy and Science 17). Leiden: Brill, 2011.

**Hirai 2012**: Hiro Hirai, "Formative Power, Soul, and Intellect in Nicoló Leoniceno between the Arabic-Latin Tradition and the Renaissance of the Greek Commentators," in Paul J.J.M. Bakker, Sander W. de Boer, and Cees Leijenhorst (eds). *Psychology and the Other Disciplines: A Case of Cross-Disciplinary Interaction (1250–1750)* (History of Science and Medicine Library 33). Leiden: Brill, 2012, pp. 297–324.

**Hissette 1977**: Roland Hissette, *Enquête sur les 219 articles condamnés à Paris le 7 mars 1277* (Philosophes Médiévaux XXII). Louvain: Publications universitaires, 1977.

**Horden 2009**: Peregrine Horden, "What's Wrong with Early Medieval Medicine," *Social History of Medicine* 24 (2009), pp. 5–25.

**Jacquart 1984/1997**: Danielle Jacquart, "De *crasis* a *complexio*: Note sur le vocabulaire du tempérament en Latin médiéval," in G. Sabbah (ed.), *Textes médicaux latins antiques*. Saint-Etienne : Université de Saint-Etienne, 1984, pp. 71–76.

Reproduced in *La science médicale occidentale entre deux renaissances (XII^e^–XV^e^ siècles)* (Variorum collected studies series 567). Aldershot, and Brookfield, VT: Variorum, 1997, no. VI.

**Jacquart 1986**: Danielle Jacquart, "A l'aube de la renaissance médicale des XIe–XIIe siècle: *L'Isagoge Johannitii* et son traducteur," *Bibliothèque de l'École des Chartes* 144 (1986), pp. 209–244.

**Jacquart 1988**: Danielle Jacquart, "Aristotelian thought in Salerno," in Peter Dronke (ed.), *A History of Twelfth-Century Western Philosophy*. Cambridge: Cambridge University Press, 1988, pp. 407–428.

**Jacquart 1993**: Danielle Jacquart, "L'influence des astres sur le corps humain chez Pietro d'Abano," in Bernard Ribémont (ed.), *Le corps et ses énigmes au Moyen Âge. Actes du colloque, Orléans, 15–16 mai 1992*. Caen : Paradigme, 1993, pp. 73–86.

**Jacquart 1994**: Danielle Jacquart, "Le sens donné par Constntin l'Africain à son oeuvre: les chapitres introductifs en arabe et en latin," in Charles Burnett and Danielle Jacquart (eds), *Constantine the African and 'Alī ibn al-'Abbās al-Magūsī: the Pantegni and Related Texts* (Studies in Ancient Medicine 10). Leiden: Brill, pp. 71–89.

**Jacquart 1993/1998**: Danielle Jacquart,"La scolastica medica", in Mirko D. Grmek (ed.), *Storia del pensiero medico occidentale*, vol. 1, *Antichità e Medioevo*. Rome-Bari: Laterza, 1993, pp. 261–322.
English translation (cited here): "Medical Scholasticism," in Mirko Grmek (ed.), *Western Medical Thought from Antiquity to the Middle Ages*. Cambridge, MA: Harvard University Press,1998, pp. 195–240.

**Jacquart 2001**: Danielle Jacquart, "Moses, Galen and Jacques Despars: Religious Orthodoxy as a Path to Unorthodox Medical Views," in Peter Biller and Joseph Ziegler (eds), *Religion and Medicine in the Middle Ages*. York: York Medieval Press, 2001, pp. 35–45.

**Jacquart 2003**: Danielle Jacquart, "Coeur ou cerveau? Les hésitations médiéevales sur l'origine de la sensation et le choix de Turisanus," in Agostino Paravicini Bagliani (ed.), *Il Cuore / The Heart* (Micrologus Library 11) Florence: SISMEL Edizioni del Galluzzo, 2003, pp. 73–95.

**Jacquart 2013**: Danielle Jacquart, "Anatomy, Physiology, and Medical Theory," in David Lindberg and Michael Shank (eds), *Cambridge History of Science*, vol. 2. *Medieval Science*. Cambridge: Cambridge University Press, 2013, pp. 590–610.

**Jacquart 2014**: Danielle Jacquart, "La complexion selon Pietro d'Abano," in *Recherches médiévales sur la nature humaine: Essais sur la réflexion médicale(XIIe–XVe s.)* (Micrologus Library 63). Florence: SISMEL Edizioni del Galuzzo, 2014, pp. 373–416.

**Jacquart 2016**: Danielle Jacquart, "Pietro d'Abano: médecin ou philosophe?," in Pieter De Leemenas and Maarten Hoenen (eds), *Between Text and Tradition: Pietro d'Abano and the Reception of Pseudo-Aristotle's* Problemata Physica *in the Middle Ages*. Leuven: Leuven University Press, 2016, pp. 1–20.

**Jacquart and Paravicini Bagliani 2007**: Danielle Jacquart and Agostino Paravicini Bagliani (eds), *La scuola medica salernitana: gli autori e i testi* (Edizione Nazionale "La Scuola Medica Salernitana" 1). Florence: SISMEL, Edizioni del Galluzzo, 2007.

**Jacquart and Micheau 1990**: Danielle Jacquart and Françoise Micheau, *La médecine arabe et l'occident médiéval* (Collection Islam-Occident). Paris: Maisonneuve et Larose, 1990.

**Jacquart and Thomasset 1985/1988**: Danielle Jacquart and Claude Thomasset, Sexualité et savoir médicale au Moyen Age. Paris: Presses Universitaires de France, 1985.
English translation (cited here): *Sexuality and Medicine in the Middle Ages*, translated by Matthew Adamson. Princeton: Princeton University Press, 1988.

**Jordan 1987**: Mark Jordan, "Medicine as Science in the Early Commentaries on Johannitius," *Traditio* 43 (1987), pp. 121–145.

**Jordan 1990**: Mark Jordan, "The Construction of a Philosophical Medicine: Exegesis and Argument in Salernitan Teaching on the Soul," *Osiris*, 2$^{nd}$ Ser. 6 (1990), pp. 42–61.

**Jordan 1992**: Mark Jordan, "The Disappearance of Galen in Thirteenth-Century Philosophy and Theology," in Albert Zimmerman and Andreas Speer (eds), *Mensch und Natur im Mittelalter* vol. 2. Berlin and New York: De Gruyter, 1992, pp. 703–717.

**Jordan 1994**: Mark Jordan, "The Fortune of Constantine's Pantegni," in Charles Burnett and Danielle Jacquart (eds), *Constantine the African and 'Alī al-'Abbās al-Maǧūsī: The* Pantegni *and Related Texts* (Studies in Ancient Medicine 10). Leiden: Brill, 1994, pp. 286–302.

**Karnes 2015**: Michelle Karnes, "Marvels in the Medieval Imagination," *Speculum* 90 (2015), pp. 327–365.

**Kelly 2001**: Donald R. Kelly, "Eclecticism and the History of Ideas," *Journal of the History of Ideas*, 62 (2001), pp. 577–592.

**Kent 1995**: Bonnie Kent, *Virtues of the Will: The Transformation of Ethics in the Late Thirteenth Century*. Washington, D.C.: Catholic University of America, 1995.

**Kibre 1978**: Pearl Kibre, "Arts and Medicine in the Universities of the Later Middle Ages," in Jozef Ijsewijn and Jacques Paquet (eds), *The Universities in the Late Middle Ages*. Leuven: Leuven University Press, 1978, pp. 213–227.

**Kieckhefer 1989**: Richard Kieckhefer, *Magic in the Middle Ages*. Cambridge: Cambridge University Press, 1989.

**Klemm 2006**: Matthew Klemm, "Medicine and Moral Virtue in the *Expositio Problematum Aristotelis* of Peter of Abano," *Early Science and Medicine* 11 (2006), pp. 302–335.

**Klemm 2012a**: Matthew Klemm, "Les complexions vertueuses: la physiologie des vertus dans l'anthropologie médicale de Pietro d'Abano," *Médiévales* 63 (2012), pp. 59–74.

**Klemm 2012b**: Matthew Klemm, "A Medical Perspective on the Soul as Form of the Body: Peter of Abano on the Reconciliation of Galen and Aristotle," in Paul J. J. M. Bakker, Sander de Boer, and Cees Leijenhorst (eds), *Psychology and the Other Disciplines: A Case of Cross-Disciplinary Interaction (1250–1750)*. Leiden: Brill, 2012, pp. 275–295.

**Klibansky, Panofsky, and Saxl 1964**: Raymond Klibansky, Erwin Panofsky, and Fritz Saxl. *Saturn and Melancholy*. London: Nelson, 1964.

**Kristeller 1945/1956**: Paul Oskar Kristeller, "The School of Salerno: Its Development and Its Contribution to the History of Learning," *Bulletin of the History of Medicine* 17 (1945), pp. 138–194 (repr. in *Studies in Renaissance Thought and Letters*, vol. 1. Rome: Storia e Letteratura, 1956, pp. 495–551).

**Kristeller 1976**: Paul Oskar Kristeller, "Bartholomaeus, Musandinus, and Maurus of Salerno and other Early Commentators of the *Articella* with a Tentative List of Texts and Manuscripts," *Italia medioevale et umanistica* 19 (1976), pp. 57–87.

**Kuksewicz 1982**: Zdzisław Kuksewicz, "The Potential and Agent Intellect," in Norman Kretzmann, Anthony Kenny, and Jan Pinborg (eds), *The Cambridge History of Later Medieval Philosophy*. Cambridge: Cambridge University Press, 1982, pp. 595–602.

**Kuksewicz 1986**: Zdzisław Kuksewicz, "A Medieval Theory of Felicity," *Dialectics and Humanism* 2–3 (1986), pp. 229–235.

**Kwakkel and Newton 2019**: Erik Kwakkel and Francis Newton with an Introduction by Eliza Glaze, *Medicine at Monte Cassino. Constantine the African and the Oldest Manuscript of his* Pantegni (Speculum Sanitatis 1). Turnhout: Brepols, 2019.

**Lagerlund 2008**: Henrik Lagerlund, "Pietro, D'Abano and the Anatomy of Perception," in Simo Knutilla and Pekka Kärkkäinen (eds), *Theories of Perception in Medieval and Early Modern Philosophy*. Dordrecht: Springer, 2008, pp. 117–130

**Lamoreaux 2016**: John C. Lamoreaux, *Ḥunayn ibn Isḥāq on His Galen Translations* (Early Christian Texts). Provo, UT: Brigham Young University Press, 2016.

**Lawn 1963**: Brian Lawn, *The Salernitan Questions: An Introduction to the History of Medieval and Renaissance Problem Literature*. Oxford: Oxford University Press, 1963.

**Lemay 1962**: Richard Lemay, *Abu Ma'shar and Latin Aristotelianism in the Twelfth Century: The Recovery of Aristotle's Natural Philosophy Through Arabic Astrology* Beirut: American University of Beirut, 1962.

**Lindberg 1983**: David Lindberg, *Roger Bacon's Philosophy of Nature. A Critical Edition, with English Translation, Introduction, and Notes, of* De multiplicatione specierum *and* De speculis comburentibus. Oxford: Oxford University Press, 1983.

**Lines 2002**: David Lines, *Aristotle's* Ethics *in the Italian Renaissance (c. 1300–1650): The Universities and the Problem of Moral Education*. Leiden: Brill, 2002.

**Livesey: 1994**: Steve Livesey, "Robert Graystanes, O.S.B. on the Subalternation of Sciences," *Recherches de théologie ancienne et médiévale* 61 (1994), pp. 136–172.

**Lloyd 1996**: Geoffrey E. R. Lloyd, *Aristotelian Explorations*. Cambridge: Cambridge University Press, 1996.

**Lloyd 2007**: Geoffrey E. R. Lloyd, "*Pneuma* between Body and Soul," *Journal of the Royal Anthropological Institute* N.S. 13 (2007), pp. S135–S146.

**Lohr 2010**: Charles Lohr, "Aristotelian 'Scientia' and the Medieval 'Artes'," in Hans-Christian Günther and Andrea A. Robiglio (eds), *The European Image of God and Man*. Leiden: Brill, 2010, pp. 259–269.

**Maccagnolo 1988**: Enzo Maccagnolo, "David of Dinant and the Beginnings of Aristotelianism in Paris," in Peter Dronke, (ed.), *A History of Twelfth-Century Western Philosophy*. Cambridge: Cambridge University Press, 1988, pp. 429–442.

**Macy 1994**: Gary Macy, "The Dogma of Transubstantiation in the Middle Ages," *Journal of Ecclesiastical History* 45 (1994), pp. 11–41.

**Mahoney 1982**: Edward Mahoney, "Sense, Intellect, and Imagination in Albert, Thomas, and Siger," in Norman Kretzmann, Anthony Kenny, and Jan Pinborg (eds), *The Cambridge History of Later Medieval Philosophy*. Cambridge: Cambridge University Press, 1982, pp. 602–622.

**Marenbon 2007**: John Marenbon, "Latin Averroism," in Anna Akasoy, James E. Montgomery and Peter E. Pormann (eds), *Islamic Crosspollinations. Interactions in the Medieval Middle East*, Exeter: Gibb Memorial Trust, 2007.

**Marengon 1984**: Paolo Marengon, "Per una revisione dell'interpretatione di Pietro d'Abano," in *Il pensiero ereticale nella Marca trevigiana e a Venezia, 1200–1350*. Abano Terme, Padua: Francisci, 1984, pp. 67–104.

**Martin 2002**: Craig Martin, "Francisco Vallés and the Renaissance Reinterpretation of Aristotle's *Meteorologica* as a Medical Text," *Early Science and Medicine* 7 (2002), pp. 1–30.

**McVaugh 1974**: Micheal R. McVaugh, "The *Humidum Radicale* in Thirteenth-Century Medicine," *Traditio* 30 (1974), pp. 259–283.

**McVaugh 1975**: Michael R. McVaugh, "Introduction" to *Arnaldi Villanova opera medica omnia* II, *Aphorismi de gradibus*. Barcelona: University of Barcelona, 1975, pp. 3–136.

**McVaugh 1990**: Michael R. McVaugh, "The Nature and Limits of Medical Certitude at Early Fourteenth-Century Montpellier," *Osiris* 2nd ser. 6 (1990), pp. 62–84.

**McVaugh 2001**: Michael R. McVaugh, "Moments of Inflection: The Careers of Arnau de Vilanova," in Peter Biller and Joseph Ziegler (eds), *Religion and Medicine in the Middle Ages*. Vol. 3. Boydell & Brewer, 2001, pp. 47–67.

**McVaugh 2003**: Michael R. McVaugh, "Incantations in Late Medieval Surgery," in Giancarlo Marchetti, Orsola Rignani, and Valeria Sorge (eds), *Ratio et superstitio: Essays in Honor of Graziella Federici Vescovini*. Turnhout: Brepols, 2003.

**McVaugh 2019**: Michael R. McVaugh, "Galen in the Medieval Universities, 1200–1400," in Petros Bouras-Vallianatos and Barbara Zipser (eds), *Brill's Companion to the Reception of Galen* (Brill's Companions to Classical Reception 17). Leiden: Brill, 2019, pp. 381–392.

**McVaugh and Siraisi 1990**: Michael R. McVaugh and Nancy Siraisi, "Renaissance Medical Learning: Evolution of a Tradition," *Osiris* 2nd ser. 6 (1990), pp. 7–15.

**Monfassani 2000**: John Monfasani "The Pseudo-Aristotelian *Problemata* and Aristotle's *De animalibus in the Renaissance*," in Anthony Grafton and Nancy Siraisi (eds), *Natural Particulars*. Cambridge, MA: MIT Press, 2000, pp. 203–247.

**Morpurgo 1984**: Piero Morpurgo, "The Salernitan School between Hippocrates, Aristotle, and Magic," *Quaderni catanesi di studi classici e medievali* 6 (1984), pp. 197–218.

**Morpurgo 1990**: Piero Morpurgo, *Filosofia della natura nella Schola Salernitana del secolo XII*. Bologna: Editrice CLUEB, 1990.

**Morpurgo 2008**: Piero Morpurgo, "I veleni nella letteratura e nell'iconografia al tempo di Pietro d'Abano," *Medicina nei Secoli* 20 (2008), pp. 525–544.

**Murdoch 1991**: John E. Murdoch, "Pierre Duhem and the History of Late Medieval Science and Philosophy in the Latin West," in Alfonso Maier and Ruedi Imbach (eds), *Gli studi di filosofia medievale fra otto e novecento*. Rome: Storia e Letteratura, pp. 253–302.

**Nardi 1912/1958**: Bruno Nardi, "La teoria dell'anima e la generazione delle forme secondo Pietro d'Abano," *Rivista di filosofia neo-scolastica* 4 (1912), pp. 723–737.
Reproduced in *Saggi sull'aristotelismo padovano dal secolo XIV al XVI*. Florence: Sansoni, 1958, pp. 1–17.

**Nardi 1920–1921/1958**: Bruno Nardi, "Intorno alle dottrine filosofiche di Pietro d'Abano," *Nuova rivista storica* 4 (1920), pp. 81–97 and 464–481; 5 (1921), pp. 300–313.
Reproduced in *Saggi sull'aristotelismo padovano dal secolo xiv al xvi*. Florence: Sansoni, 1958, pp. 19–74.

**Newton 1994**: Francis Newton, "Constantine the African and Monte Cassino: New Elements and the Text of the *Isagoge*," in Charles Burnett and Danielle Jacquart (eds), *Constantine the African and 'Alī al-'Abbās al-Maǧūsī: The* Pantegni *and Related Texts* (Studies in Ancient Medicine 10). Leiden: Brill, 1994, pp. 16–47.

**Normore 1995**: Calvin Normore, "Who Was Condemned in 1277?" *The Modern Schoolman*, 72 (1995), pp. 273–281.

**Nutton 2020**: Vivian Nutton, *Galen: A Thinking Doctor in Imperial Rome* (Routledge Ancient Biogfraphies). London and New York, NY: Routledge, 2020.

**Nutton with Bos 2015**: Vivian Nutton with Gerrit Bos, *On Problematical Motions*. Cambridge: Cambridge University Press, 2015.

**O'Boyle 1992**: Cornelius O'Boyle, "Medicine, God and Aristotle in the Early Universities: Prefatory Prayers in Late Medieval Medical Commentaries," *Bulletin of the History of Medicine* 66 (1992), pp. 185–209.

**O'Boyle 1998**: Cornelius O'Boyle, *The Art of Medicine: Medical Teaching at the University of Paris, 1250–1400* (Education and Society in the Middle Ages and Renaissance 9). Leiden: Brill, 1998.

**Ottaviani 2004**: Didier Ottaviani, "Le méthode scientifique dans le *Conciliator* de Pietro d'Abano," in Christophe Grellard, (ed.), *Méthodes et statut des sciences à la fin du Moyen Age*. Lille: Presses Universitaires du Septentrion, 2004, pp. 13–26.

**Ottosson 1984**: Per-Gunnar Ottosson, *Scholastic Medicine and Philosophy: A Study of Commentaries on Galen's Tegni (ca. 1300–1450)*. Naples: Bibliopolis, 1984.

**Page 2013**: Sophie Page, *Magic in the Cloister: Pious Motives, Illicit Interests, and Occult Approaches to the Medieval Universe*. University Park, PA: Penn State University Press, 2013.

**Park 1988**: Katharine Park, "Psychology," in *Cambridge History of Renaissance Philosophy*. Cambridge: Cambridge University Press, 1988, pp. 464–484.

**Paschetto 1984**: Eugenia Paschetto, *Pietro d'Abano. Medico e filosofo*. Florence: Vallecchi, 1984.

**Pasnau 2002**: Robert Pasnau, *Thomas Aquinas on Human Nature*. Cambridge: Cambridge University Press, 2002.

**Pasnau 2012a**: Robert Pasnau, "The Latin Aristotle," in Christopher Shields (ed.), *The Oxford Handbook of Aristotle*. Oxford: Oxford University Press, 2012, pp. 665–690.

**Pasnau 2012b**: Robert Pasnau, "Mind and Hylomorphism," in John Marenbon (ed.), *The Oxford Handbook of Medieval Philosophy*. Oxford: Oxford University Press, 2012, pp. 486–504.

**Pasnau and Van Dyke 2010**: Robert Pasnau and Christina Van Dyke (eds), *The Cambridge History of Medieval Philosophy*, 2 vols. Cambridge: Cambridge University Press, 2010.

**Pellegrin 1982/1986**: Pierre Pellegrin, *Classification des Animaux chez Aristote. Statut de la biologie et unité de l'aristotélisme* (Collection d'études anciennes). Paris: Belles Lettres, 1982.
English translation (cited here): *Aristotle's Classification of Animals: Biology and the Conceptual Unity of the Aristotelian Corpus*. Transl. Anthony Preus. Berkeley, CA: University of California Press, 1986.

**Pesenti 1980**: Tiziana Pesenti, "Per la Tradizione del Testamento di Pietro D'Abano," *Medioevo* 6 (1980), pp. 533–542.

**Pesenti 2001**: Tiziana Pesenti, "The Libri Galieni in Italian Universities in the Fourteenth Century," *Italia medioevale e umanistica* 42 (2001), pp. 119–147.

**Pink 2012**: Thomas Pink, "Freedom of the Will," in John Marenbon (ed.), *The Oxford Handbook of Medieval Philosophy*. Oxford: Oxford University Press, 2012, pp. 569–587.

**Prioreschi 1996**: Plinio Prioreschi. "Galenicae Queaestiones Disputatae Duae: Rete Mirabile and Pulmonary Circulation," *Vesalius* 2 (1996), pp. 67–78.

**Randall 1940/1961**: John Herman, Randall Jr., "The Development of Scientific Method in the School of Padua," *Journal of the History of Ideas* 1 (1940), pp. 177–206.
Reproduced in *The School of Padua and the Emergence of Modern Science*. Padua: Antenore, 1961.

**Renan 1852**: Ernest Renan, *Averroès et averroïsme*, Paris: Auguste Durand, 1852.

**Robert 2014**: Aurélien Robert, "Le pouvoir des incantations selon les médecins du Moyen Âge (XIIIe–XVe siècle)," in Nicole Bériou, Jean-Patrice Boudet, et Irène Rosier-Catach (eds), *Le pouvoir des mots au Moyen Âge*. Turnhout, 2014, pp. 459–489.

**Robert 2017**: Aurélien Robert, "Pietro d'Abano et matérialisme," in Tiziana Suarez-Nani and Agostino Paravicini Bagliani (eds), *Materia. Nouvelles perspectives de recherche dans la pensée et la culture médiévales (XIIe–XVIe siècles)* (Micrologus Library 83). Florence: SISMEL Edizioni del Galluzzo, 2017, pp. 217–251.

**Robert 2023**: Aurélien Robert, "The Concept of Complexion in Antonio da Parma's Medical Anthropology," *Early Science and Medicine* 28 (2023), pp. 295–354.

**Ronca 1994**: Italo Ronca, "The Influence of the *Pantegni* on William of Conches' *Dragmaticon*," in Charles Burnett and Danielle Jaquart (eds), *Constantine the African and 'Alī al-'Abbās al-Mağūsī: The Pantegni and Related Texts* (Studies in Ancient Medicine 10) Leiden: Brill, 1994, pp. 266–287.

**Saarinen 1994**: Risto Saarinen, *Weakness of the Will in Medieval Thought: From Augustine to Buridan*. Leiden: Brill, 1994.

**Saffron 1972**: Morris Harold Saffron, *Maurus of Salerno, Twelfth-Century "Optimus Physicus" with his Commentary on the* Prognostics *of Hippocrates*. Philadelphia, PA: American Philosophical Society, 1972.

**Salmón 2012**: Salmón, Fernando, "The Physician as Cure in Medical Scholasticism," in Ildikó Csepregi and Charles Burnett (eds), *Ritual Healing: Magic, Ritual and Medical Therapy from Antiquity until the Early Modern Period* (Micrologus Library 48). Florence: SISMEL Edizioni del Galluzzo, 2012, pp. 193–215.

**Schmitt 1985**: Charles B. Schmitt, "Aristotle Among the Physicians" in Andrew Wear, Roger K. French, and I. M. Lonie (eds), *The Medical Renaissance of the Sixteenth Century*, Cambridge: Cambridge University Press, 1985, pp. 1–15.

**Sears 1986**: Elizabeth Sears, *The Ages of Man: Medieval Interpretations of the Life Cycle*. Princeton: Princeton University Press, 1986.

**Sela 2019**: Shlomo Sela, "Pietro d'Abano, Translator of Abraham Ibn Ezra's Astrological Writings," *Sefarad* 79 (2019), pp. 7–87.

**Sigerist 1958**: Henry Sigerist. "The Latin Medical Literature of the Early Middle Ages," *Journal of the History of Medicine and Allied Sciences* 13 (1958), pp. 127–143.

**Siraisi 1970**: Nancy Siraisi, "The *Expositio Problematum Aristotelis* of Peter of Abano," *Isis* 61 (1970), pp. 321–339.

**Siraisi 1973**: Nancy Siraisi, *Arts and Sciences at Padua: The* studium *of Padua before 1350*. Toronto: Pontifical Institute of Medieval Studies, 1973.

**Siraisi 1976**: Nancy Siraisi, "The *Libri Morales* in the Faculty of Art and Medicine at Bologna: Bartolomeo da Varignana and the Pseudo-Aristotelian *Economics*," *Manuscripta* 20 (1976), pp. 105–118.

**Siraisi 1981**: Nancy Siraisi, *Taddeo Alderotti and His Pupils: Two Generations of Medical Learning*. Princeton: Princeton University Press, 1981.

**Siraisi 1985**: Nancy Siraisi, "Pietro d'Abano and Taddeo Alderroti: Two Models of Medical Culture," *Medioevo* 11 (1985), pp. 139–162.

**Siraisi 1987**: Nancy Siraisi, *Avicenna in Renaissance Italy: The* Canon *and Medical Teaching in Italian Universities after 1500*. Princeton: Princeton University Press: 1987.

**Siraisi 1990**: Nancy Siriasi, *Medieval and Early Renaissance Medicine: An Introduction to Knowledge and Practice*. Chicago: University of Chicago Press, 1990.

**Siraisi 1992**: Nancy Siraisi, "The Faculty of Medicine," in Hilde De Ridder-Symoens (ed.), *A History of the University in Europe*, vol. 1: *Universities in the Middle Ages*. Cambridge: Cambridge University Press, 1992, pp. 360–387.

**Southern 1982**: Richard W. Southern, "The Schools of Paris and the School of Chartres," in Robert Benson and Giles Constable (eds), *Renaissance and Renewal in the Twelfth Century*. Cambridge, MA: Harvard University Press, 1982, pp. 113–137.

**Steneck 1974**: Nicholas H. Steneck, "Albert the Great on the Classification and Localization of the Internal Senses," *Isis* 65 (1974), pp. 193–211.

**Stock 1972**: Brian Stock, *Myth and Science in the Twelfth Century: A Study of Bernard of Sylvester*. Princeton: Princeton University Press, 1972.

**Struever 1993**: Nancy Struever, "Petrarch's *Invective contra medicum*: An Early Confrontation of Rhetoric and Medicine," *Modern Language News* 108 (1993), pp. 659–679.

**Swain 2007**: Simon Swain, (ed.), *Seeing the Face, Seeing the Soul: Polemon's Physiognomy from Classical Antiquity to Medieval Islam*. Oxford: Oxford University Press, 2007.

**Tachau 1988**: Katherine Tachau, *Vision and Certitude in the Age of Ockham: Optics, Epistemology, and the Foundation of Semantics*. Brill: Leiden, 1988.

**Tachau 1993**: Katherine Tachau, "What Senses and Intellect Do: Argument and Judgment in Late Medieval Theories of Knowledge," in Klaus Jacobi (ed.), *Argumentationstheorie: Scholastische Forschungen zu den logischen und semantischen Regeln korrekten Folgerns*. Leiden: Brill, 1993.

**Tachau 1997**: Katherine Tachau, "*Et maxime visus, cuius species venit ad stellas et ad quem species stellarum veniunt*: Perspectiva and Astrologia in Late Medieval Thought," in Agostino Paravicini Bagliani (ed.), *La Visione e lo sguardo nel Medio Evo* (Micrologus Library 5). Florence: SISMEL Edizioni del Galluzzo, 1997, pp. 201–224.

**Tachau 1999**: Katherine Tachau, "Some Aspects of the Notion of Intentional Existence at Paris, 1250–1320," in Sten Ebbesen and Russell Friedman (eds), *Medieval Analyses in Language and Cognition*, Copenhagen: Royal Danish Academy of Sciences and Letters, 1999, pp. 331–354.

**Takahashi 2008**: Adam Takahashi, "Nature, Formative Power, and Intellect in the Philosophy of Albert the Great." *Early Science and Medicine* (2008), pp. 439–458.

**Temkin 1949**: Owsei Temkin, "Medicine and the Problem of Moral Responsibility," *Bulletin of the History of Medicine* 23 (1949), pp. 1–20.

**Temkin 1951**: Owsei Temkin, "On Galen's Pneumatology," *Gesnerus* 8 (1951), pp. 180–189.

**Temkin 1973**: Owsei Temkin, *Galenism: Rise and Decline of a Medical Philosophy*. Ithaca, NY: Cornell University Press, 1973.

**Temkin 1991**: Owsei Temkin, *Hippocrates in a World of Pagans and Christians*. Baltimore, MD, and London: Johns Hopkins University Press, 1991.

**Thijssen 1991**: Hans Thijssen, "Some Reflections on Continuity and Transformation of Aristotelianism in in Medieval (and Renaissance) Natural Philosophy," *Documenti e studi sulla tradizione filosofica medievale* 2 (1991), pp. 503–528

**Thijssen 1997**: Hans Thijssen, "1277 Revisited: A New Interpretation of the Doctrinal Investigations of Thomas Aquinas and Giles of Rome," *Vivarium* 35 (1997), pp. 72–100.

**Thijssen 1998**: Hans Thijssen, *Censure and Heresy at the University of Paris, 1200–1400*. Philadelphia, PA: University of Pennsylvania Press, 1998.

**Thijssen 2023**: Hans Thijssen, "Condemnation of 1277," in Edward N. Zalta and Uri Nodelman (eds), *The Stanford Encyclopedia of Philosophy* (Winter 2023 Edition). https://plato.stanford.edu/archives/win2023/entries/condemnation/ (last access 07.08.2024).

**Thomann 1991**: Johannes Thomann. "Pietro d'Abano on Giotto," *Journal of the Warburg and Courtauld Institutes* 54 (1991), pp. 238–244.

**Thorndike 1923**: Lynn Thorndike, *A History of Magic and Experimental Science*, vol. 2 *During the First Thirteen Centuries of Our Era*. New York, NY: Columbia University Press, 1923.

**Thorndike 1926**: Lynn Thorndike, "Relations of the Inquisition of Peter of Abano and Cecco d'Ascoli." *Speculum* 1 (1926), pp. 338–343.

**Thorndike 1942**: Lynn Thorndike, "Translations of the Works of Galen from the Greek by Peter of Abano," *Isis* 33 (1942), pp. 649–653.

**Thorndike 1944a**: Lynn Thorndike, "Manuscripts of the Writings of Peter of Abano," *Bulletin of the History of Medicine* 15 (1944), pp. 201–219.

**Thorndike 1944b**: Lynn Thorndike, "The Latin Translations of the Astrological Texts of Abraham Avenezra," *Isis* 35 (1944), pp. 293–302.

**Thorndike 1958**: Lynn Thorndike, "De Complexionibus," *Isis* 49 (1958), pp. 398–408.

**Touwaide 2008**: Alain Touwaide, "Pietro d'Abano sui veleni. Tradizione medievale e fonti greche," *Medicina nei Secoli* 20 (2008), pp. 591–605.

**Touwaide 2024**: Alain Touwaide, "Physiognomy: The Origins," in Massimo Ciavolella, Valeria Finucci, and Megan Tomlison (eds), *Physiognomy at the Crossroad of Magic, Science and the Arts* (Medical Traditions 11). Berlin and Boston: De Gruyter, 2024, pp. 23–33.

**Van der Lugt 2013**: Maaike van der Lugt, "The Learned Physician as Charismatic Healer: Urso of Salerno (Flourished End of Twelfth Century) on Incantations in Medicine, Magic, and Religion," *Bulletin for the History of Medicine* 87 (2013), pp. 307–346.

**Van Oppenraay 2017**: Aafke M. I. Van Oppenraay, "Avicenna's *Liber de animalibus* (*'Abbreviatio Avicennae'*). Preliminaries and State of Affairs," *Documenti e studi sulla tradizione filosofica medievale* 28 (2017), pp. 401–416.

**Van Steenberghen 1970**: Fernand Van Steenberghen, *Aristotle in the West: The Origins of Latin Aristotelianism*, 2nd ed., transl. Leonard Johnson. Louvain: Nauwelaerts, 1970.

**Van Steenberghen 1980**: Fernand Van Steenberghen, *Thomas Aquinas and Radical Aristotelianism*, Washington, D.C: Catholic University of America Press, 1980.

**Van Steenberghen 1991**: Fernand Van Steenberghen *La philosophie au XIIIe siècle*. Louvain: Institut supérieur de philosophie, and Leuven: Peeters, 1991.

**Vasoli 1974**: Cesare Vasoli, "Per un ricognizione delle fonti della storia della scienza in Italia," *Profezia e Ragione: Studi sulla cultura del Cinquecento e del Seicento*. Naples: Morano, 1974, pp. 405–475.

**Ventura 2016**: Iolanda Ventura, "Botany, Dietetics, and Pharmacy in Pietro d'Abano's Expositio Problematum: On Sections XX, XXI, and XXII," in Pieter De Leemeans and Maarten Hoenen (eds), *Between Text and Tradition: Pietro d'Abano and the Reception of Pseudo-Aristotle's* Problemata Physica *in the Middle Ages*. Leuven: Leuven University Press, 2016, pp. 163–200.

**Vescovini 1985**: Graziella Federici Vescovini, "Pietro d'Abano e le fonti astronomiche greco-arabo-latine (a propositio del *Lucidator astronomiae* o *astrologiae*)," *Medioevo* 11 (1985), pp. 65–96.

**Vescovini 1987**: Graziella Federici Vescovini, "Peter of Abano and Astrology," in Patrick Curry (ed.), *Astrology, Science, and Society: Historical Essays*. Woodbridge: Boydell and Brewer, 1987, pp. 19–39.

**Vescovini 1997**: Graziella Federici Vescovini, "L'antropologia naturale di Pietro d'Abano," *Paradigmi* 45 (1997), pp. 525–541.

**Vescovini 1999**: Graziella Federici Vescovini, "La medicine synthese d'art et de science selon Pierre d'Abano," in Roshdi, Rashed and Joel Biard (eds), *Les doctrines de la science de l'antiquite a l'âge classique*. Leuven: Peeters, 1999, pp. 238–255.

**Vescovini 2006**: Graziella Federici Vescovini, "*L'Exposito Succinta Problematum Aristotelis* de Pierre d'Abano," in Pieter De Leemans and Michèle Goyens (eds), *Aristotle's* Problemata *in Different Times and Tongues*. Leuven: Leuven University Press, 2006, pp. 55–70.

**Wack 1990**: Mary F.Wack, *Lovesickness in the Middle Ages: The "Viaticum" and its Commentaries*. Philadelphia, PA: University of Pennsylvania Press, 1990.

**Wallace 1995**: William Wallace, "Circularity and the Paduan *Regressus*: From Pietro d'Abano to Galileo Galilei," *Vivarium* 33 (1995), pp. 76–97.

**Wallis 1995**: Faith Wallis, "The Experience of the Book: Manuscripts, Texts, and the Role of Epistemology in Early Medieval Medicine," in Don Bates (ed.), *Knowledge and the Scholarly Traditions*. Cambridge: Cambridge University Press, 1995, pp. 101–126.

**Wallis 2000**: Faith Wallis, "Signs and Senses: Diagnosis and Prognosis in Early Medieval Pulse and Urine Texts," *Social History of Medicine* 13 (2000), pp. 265–278.

**Wallis 2007**: Faith Wallis, "The Articella Commentaries of Master Bartholomaeus," in Agostino Paravicini Bagliani (ed.), *La Scuola medica salernitana: gli autori e i testi* (Edizione Nazionale "La Scuola Medica Salernitana" 1). Florence: SISMEL Edizioni del Galluzzo, 2007, pp. 125–164.

**Wallis 2008**: Faith Wallis, "Twelfth-century Commentaries on the Tegni: Bartholomaeus of Salerno and Others," in Nicoletta Palmieri (ed.), *Les parcours de l'Ars medica (Tegni) de Galien: lectures et interpétations depuis la fin de l'Antiquité jusqu'aux Universités médiévales*. Saint-Étienne: Centre Jean-Palerne, 2008, pp. 127–168.

**Wallis 2010**: Faith Wallis, *Medieval Medicine: A Reader*. Toronto: University of Toronto Press, 2010.

**Wallis 2015**: Faith Wallis, "Character and Complexion in Twelfth-Century Medicine," in Hélène Cazes and Anne-France Morand (eds), *Miroirs de la mélancholie*. Paris: Hermann, 2015, pp. 59–86.

**Weill-Parot 2010**: Nicolas Weill-Parot, "Astrology, Astral Influences, and Occult Properties in the Thirteenth and Fourteenth Centuries," *Traditio* 65 (2010), pp. 201–230.

**Weisheipl 1978**: James A. Weisheipl, "The Nature, Scope, and Classification of the Sciences," in David Lindberg (ed.), *Science in the Middle Ages*. Chicago: University of Chicago Press, 1978, pp. 461–482.

**Wetherbee 1988**: Winthrop Wetherbee, "Philosophy, Cosmology, and the Twelfth–Century Renaissance," in Peter Dronke (ed.), *A History of Twelfth–Century Western Philosophy*. Cambridge: Cambridge University Press, 1988, pp. 21–53.

**Wicher 1986**: Helen Brown Wicher, "Nemesius Emesenus," in Paul Oskar Kristeller and F. Edward Cranz (eds), *Catalogus Translationum et Commentariorum*, vol. 6. Washington, D.C.: Catholic University of America Press, 1960, pp. 31–68.

**Wilcox and Riddle 1995**: Judith Wilcox and John M. Riddle, "Questa ibn Luqa's *Physical Ligatures* and the Recognition of the Placebo Effect," *Medieval Encounters* 1 (1995), pp. 1–50.

**Wippel 1977**: John Wippel, "The Condemnations of 1270 and 1277 at Paris," *Journal of Medieval and Renaissance Studies* 7 (1977), pp. 169–201.

**Wippel 1995**: John Wippel, "Thomas Aquinas and the Condemnation of 1277," *The Modern Schoolman* 72 (1995), pp. 233–273.

**Wolfson 1935**: Harry Austryn Wolfson, "The Internal Senses in Latin, Arabic and Hebrew Philosophical Texts," *Harvard Theological Review* 28 (1935), pp. 69–133.

**Ziegler 2008**: Joseph Ziegler, Review of Jacquart and Paravicini Bagliani 2007, *Aestematio* 5 (2008), pp. 81–92.

**Zupko 1997**: Jack Zupko, "What is Science of the Soul? A Case Study in the Evolution of Late Medieval Natural Philosophy," *Synthèse* 110 (1997), pp. 297–334.

# General Index

The index below does not include terms that are ubiquitous in this book. These include *body, soul, virtue* (along with its translations as *power, faculty,* or *capacity*), *cause, intellect, nature,* and other common terms. However, many individual virtues (e.g., estimation) are included. Likewise, this index does not include Aristotle or Galen, but does include their individual works.


'Ali ibn al-'Abbās (*see* Haly Abbas)
Abraham ibn Ezra 7
Abstraction, process of 77, 80, 85, 90, 97
Abu Ma'shar 24
Action at a distance 93
Actuality 50, 67, 71, 75
Adam and Eve, natures of 26
Adelard of Bath 25, 29
Affections of body and soul 43, 64, 82, 88 (*see also* emotions and passions)
Age 35, 48, 52, 97
– stages of life 91–92
Air 20, 24, 48, 65
Albertus Magnus 3, 27, 32, 36, 62, 77n6, 101
– *De animalibus* 40
– *Quaestiones de animalibus* 40
Alchemists 63
Alexander of Aphrodisias 84, 92
– *Problems* 12n68
Alexandria 9, 17n5, 18, 37
– Alexandrian developments 17n4
– Alexandrian Galenism 19
Alfanus of Salerno 18
– translator of Nemesius 21
Alfred of Sareshel 27
*Almansor* (see al-Rāzī)
al-Rāzī (Rhazes or Rasis) 14
– *Almansor* 32
Angels 60, 93, 105
Anger 82, 98, 116, 117, 125
Animal capacities, faculties, powers, or virtues 19, 22, 55n49, 67, 78, 81, 82, 96, 106, 124, 125n70
– animal spirits 20, 27, 61, 62, 63
Animals 6, 14, 34–36, 47–49, 52, 54, 59, 63, 64, 78–81, 87, 88, 90, 98, 117, 118, 122, 124, 125
– beasts 121
– bestial excesses 59
– bestial habits and appearance 87
– bestial passions 125
– bestial pleasures 121
Anselm of Bec 25
Anthropology, Christian 1, 21, 26, 28
*Aphorisms* (Urso of Salerno) 24
*Aphorisms*, Hippocratic 18n10, 25, 94, 95n13, 131
– Galen on the Hippocratic *Aphorisms* 22n30, 33, 38
Appetitive power, faculty, or virtue 77, 78
Apprehension 77, 97 (*see also* perception, sensation, and senses)
Arabic language 20, 25, 32, 34n119, 35n122
– Arabic Galenism 17, 61
– Arabic philosophy 34
– Arabic physicians 7, 49
– Arabic texts or authors 2, 6, 14, 17, 18, 19, 21, 24, 32, 37, 40
Archimatthaeus 22
Aristotelianism(s) 5–6
Arnau of Vilanova 37, 38, 40, 41, 101
*Ars commentata* 33
*Ars notoria* 102
*Ars Parva* (*see Tegni*)
Art (vs. science) 1n1, 6, 31, 32n98, 38 (*see also* mechanical arts)
– arts and melancholy 59
– liberal arts 20, 112
– practical arts 32
Arteries 19, 27, 62, 63, 65, 70, 82
Arthritis 131
*Articella* 18, 22, 32, 33, 36, 94
Arts faculty at Paris 3, 34, 35, 36, 42, 43, 113
Astrology 1, 3n12, 4, 6, 7, 9, 14 (*see also*, celestial and planets)
– astral forces 6
– astral magic 101
– astrologers 8
– astrological doctrine 5
– astrological determinism 43
– astrological texts 7, 24
Augmentative powers 77, 78
Augustine 41
– *De Genesi ad litteram* 22


https://doi.org/10.1515/9783110612639-010

Averroes 42, 49, 50, 51, 53, 66, 75, 79, 84, 92, 108, 122n52
– *Colliget* 40, 47, 49
– *Long Commentary on the* De anima *of Aristotle* 34n119
Averroism 4, 84
– Averroist theory of the unity of the intellect 42, 84, 86
Avicenna 9, 14, 37, 41, 49, 55n47, 57, 60n71, 61, 62, 66, 67n38, 77, 79, 80, 86, 88, 106, 108, 109, 124
– *Canon* 9, 32, 33, 37, 38, 40, 41n157, 45n1, 45n7, 47, 61n2, 76n4, 91n70, 127n79
– *De anima* 34, 53n37, 62n12, 77, 79n10, 79n13, 85, 86, 88m58, 95, 96n23, 108, 127n79
– *De animalibus* 40
– *Metaphysica* 104n60, 105n65

Baghdad 18, 19, 20
Balance 19, 29, 49, 60, 96, 111, 122, 123, 126, 131 (*see also* imbalance and temperance)
– balance of humors, qualities, or spirits 45, 46, 51, 57–60, 115, 124, 129
– morally balanced 23
– perfect balance 50, 51, 57–60
Bartholomew of Messina 12–13, 14n72
Bartholomew of Salerno 23
Bartolomeo da Varignana 39
Beatitude of the soul (*see* perfection)
Behavior 14, 20, 52, 114, 122n56
– illicit behavior 120
– virtuous behavior 129
Bernard of Clairvaux 29
Bernard Silvester 26
Bible 30
– Biblical exegesis 29
– *Song of Songs* 29
Bile 1, 115
– black bile 59 (*see also* melancholy)
Blood 1, 14, 47, 48, 52, 65, 68, 70, 74, 77, 82, 103, 116, 118, 119
Boethius, on Aristotle's *Topics* 22n30
Boldness 49 (*see also* courage and brave)
Bologna 3, 33n106, 34, 36–38
Bone 47, 48, 103
Bono, James 27, 62, 74
Bowels 115–116
Brain 19–23, 27, 32, 40, 48, 53, 62, 63, 64, 65, 67, 68, 81, 82, 103
Brave or bravery 14, 116, 118, 119 (*see also* boldness and courage)
Burgundio of Pisa 21n22, 32

Cadden, Joan 132
Calcidius 26
*Canon* (*see* Avicenna)
*Categoriae* (Aristotle) 24
Cathars or Albigensians 30
Celestial bodies 6–8, 42, 46, 58, 68
– celestial causes 43
– celestial forces or powers 14, 71, 73, 93, 131
– celestial heat 61, 65, 69, 74
– celestial substance 27
– eternal celestial motion 34
– celestial qualities 62
– celestial form 69
– spheres 6
– stars 7, 10, 132
– heavens 7, 69, 73n5, 105
Censures 28, 35, 36
Cerebrum (*see* brain)
Character 15
– characters for use in spells 103
– moral character 23
Charms 94
Chartres 21, 22n26, 25, 26
Choice 117–119
Choleric complexion 58
Christ (*see* Jesus)
Clergy 28, 30
Cogitation or cogitative power 19, 22, 53, 78, 79, 88, 99
Coitus 68
Cologne 40
Common sense 78, 79, 97
*Compilatio Physionomiae* 3n12, 9, 92, 98
description of 14–15
Concept 80
Conception 64
*Conciliator*, description and structure of 9–10
Concupiscible power 77, 78, 82, 113, 119–123, 124, 128
– concupiscible excesses 59
Condemnation of 1277 (and 1270) 8n49, 42–43, 108, 119, 129n91 (*see also* censure)
Conjunctions of the planets 60, 131
Conservative power 78, 80
Constantine the African 18–20, 21, 26, 27n69, 103
– *De coitu* 29

Constantinople 3, 12n68
Continence 113, 119, 126
Corruption, moral 25, 43
*Cosmographia* (*see* Bernard Silvester)
Cosmos 25, 26, 58
Costa ben Luca (*see* Qustā ibn Lūqā)
Courage 23, 52, 113, 115–119
*Crises and critical days* (Galen) 10
Curriculum, scholastic 13
– arts curriculum 36
– medical curriculum 6, 18, 24n45, 33, 94

David of Dinant, *Quaternuli* 35
*De accidenti et morbo* (Galen) 117
*De anima* (Aristotle) 35, 59n68, 66, 67, 73n67, 77n8, 80, 83n32, 88, 89, 91n71, 97, 98, 99, 121n51, 127n79 (for Avicenna's *De Anima,see* Avicenna)
*De animalibus* (*see Historia animalium*, *De partibus animalium*, and *De generatione animalium*)
*De coitu* (see Constantine the African)
*De complexionibus* (Galen) 33, 47, 48, 49n20–21, 57n56
*De differentia anime et spiritus* (see Qustā ibn Lūqā)
*De generatione animalium* (Aristotle) 35, 72n62, 73, 89n62, 90
*De generatione et corruptione* (Aristotle) 24, 72n63
*De interioribus* (Galen) 33
De Leemans, Pieter 3n9, 13
*De memoria et reminiscentia* (Aristotle) 80
*De motu animalium* (Aristotle) 88, 93, 98n38, 102
*De motu octavae spherae* 7
*De natura hominis* (*see* Nemesius)
*De optima compositione* (Galen) 55
*De partibus animalium* (Aristotle) 1n1, 35, 47, 48n17, 52n32–34
*De philosophia mundi* (see William of Conches)
*De tremore, iectigatione, et spasmo* (Galen) 53
*De utilitate partium* (Galen) 55
Deficiencies, natural 44
Delaurenti, Béatrice 107, 109
Demons 93, 105, 107
– demonic favor and depravity 63
Desire 21, 24, 77, 121, 126–128
– bodily 28
– desire for illicit pleasures 120
– desire for wine 118
– sexual desire 43, 121
Dexterity 20
Diets 33, 46, 66, 129, 131
Digby commentary 22
Digestion 10, 19, 65
Dino del Garbo 71
Disease 6, 28, 29, 52
– illness 8, 9n52, 10, 19–21, 28, 29, 95, 99, 100
– sick 7, 10, 23, 30, 95, 96, 108, 125
– sickness 1, 28, 30, 101, 126
Dispositions 14, 19, 30, 74, 86, 96, 97, 109, 116, 122n52, 131
Divine 7, 8, 9n52, 27, 30, 33, 60, 62, 68, 69n50, 71, 73, 74, 77, 90, 94
– derogating divine wisdom 3n12
– divine illumination 8
– divine omnipotence 108
– divine truth 25
– divinely infused 84
Doctrine, Christian 22, 28, 29n81, 31, 42, 53
Dogs 48, 101, 103
Dominicans 3, 7, 36, 73, 83
*Dragmaticon* (*see* William of Conches)
Drink (*see* food and drink)

Ears 14
Effective power 53, 77, 78
Embryo 61, 64, 89
– animation of the embryo 67–73
Emotions 19, 20, 77, 94, 97, 113 (*see also* affections and passions)
Envy or envious 15, 49
Equality, of qualities 58 (*see also* balance)
– inequality 119 (*see also* imbalance)
Estimation or estimative power 78–80, 97, 99, 104
*Ethica Nicomachea* (Aristotle) 5n33, 32n99, 34, 35n123, 43, 44, 113n11, 114n14, 119, 123n58, 124, 129
Ethics 23, 52n30, 78, 80, 113
– elitist ethical ideal 43
– ethical discussions 113n12, 129
Eucharist 30, 102
Excesses 59, 87, 127, 128
*Expositio Problematum Aristotelis*, description and structure of 11–13
Eyes 15, 82, 116

Face 15, 116
Faculties at the schools 3
– medical faculties 31, 42
Fate 21

Fear 13n69, 13n70, 49, 82, 93, 98, 115–118, 127
Females 14, 71, 131
Fetus 71
Fevers 10
Flesh 47, 48, 55, 63
Food and drink 20, 51, 65, 121, 122n56, 131 (*see also* wine)
Form, substantial 46, 53–57, 60, 67, 70–73, 81, 84
Fortune 8
Free will 21, 83
French (language) 7
Friendly 49

Galenism 17, 19, 40, 61
Garcia-Ballester, Luis 54n43
al-Gazel 108
Generation 21, 48, 65
– generation of the soul 67–73
Gentile da Cingoli 95
Gerard of Cremona 14n75, 32
Giotto vii, 15n84
Gluttony 121, 131
God 7, 8, 27, 28, 30, 33, 39, 42, 45, 57, 68, 74, 89, 91, 105, 128
Greek language 3, 12, 18, 21, 32, 34n119, 35n121, 35,n122
– texts or sources 2, 11n62, 12, 14, 17, 18, 21, 55
Gregory the Great (*see* Pope)
Growth 19, 21, 69, 70, 71, 77, 81 (*see also* augmentative)

Habits 14, 46, 87, 113, 118–120, 122, 124, 126, 127, 128
– habituation of the body 111, 119, 126
Hairy 14
Haly Abbas 9, 20, 45n7, 47, 66, 124
Hands 45, 89, 91, 115, 116
Happiness 43 44, 113, 114, 132
Hasse, Dag 84
Health 1, 8, 10, 17, 19–21, 23, 29, 40, 45, 46, 52, 60, 81, 101, 112, 126, 131
– confidence and health 95–100
– health of the soul 28–29
– healthy amount of heat 23, 119
– healthy complexions 58
– healthy operation of the lower virtues 91
Heart 19, 20, 23, 52, 99
– heart and origin of life 67–70
– heart and spread of emotions 77, 115–118
– heart and vital spirits 27, 62–66, 82, 99
– primacy of the heart 32, 40, 67, 81 (*see also* parts or organs, principal)
Heresy 3, 30, 36
Hippocrates 28, 41 (*see also Aphorisms* and *Prognostics*)
– Hippocratic literature 18
*Historia animalium* (Aristotle) 35
Hugh of St. Victor, *Didascalicon* 29
Hylomorphism 1, 65, 71, 83

Ibn Ishāq, Hunayn (*see* Johannitius)
Ibn Sina (*see* Avicenna)
*Iliad* 101
Illness (*see* disease)
Imagination 1, 19, 21–22, 75, 78, 79, 84, 88, 89, 91, 93, 99, 100, 103
– corruption of the imagination or interior senses 88, 91
Imbalance 28, 59, 60, 118, 125
Immortality of the soul 21, 22, 35
Inanimate things 21
Inclinations 23, 26, 118
Incontinence 59, 119–120
– incontinent people 120, 129
Infirm (or *infirmitas*) 28, 30
Infirmaries, monastic 30
Informative virtue 68–75, 77, 89–91
– corruption of the informative virtue 71
Innocent III (see Pope)
Instincts 79, 118
*Intellectus vocatus* 3n12, 72–73, 75, 90
Intelligences, separated 7, 8, 39, 86, 93, 104, 105, 106
Intentionality 27
Intentions or intentional species 7, 78, 80, 95n20, 96–100, 106 (*see also* species)
Internal, interior, or occult senses 8, 78–80, 88, 91, 96, 97, 106
*Invective contra medicum* (*see* Petrarch)
Involuntary behavior 82, 116–118
Irascible vice 59, 121
– irascible desires 124, 128
– irascible habits 113, 115–118
– irascible power 77, 78, 82, 124
Isaac Israeli, *Universal Diets* 22n30
*Isagoge* (*see* Johannitius)
Isidore of Seville, *Etymologies* 17

Jacquart, Danielle 2n2, 3n12
James of Venice 34n119, 35n121
Jean de Jandun 11n59–60
Jesus (or Christ) 30, 33, 60
– nature of Jesus 9, 26
Johannitius 19
– *Isagoge* 9, 18, 18n10, 19, 20, 22, 23, 25, 29, 37, 47, 55n47, 62
Jordan, Mark 22
Joy 82, 97, 117
Justice 113, 122–129
– complexion of justice 58
Justinian, *Institutes* 123

Latin Averroism (see Averroism)
Liberal arts 20, 112
*Libri naturales* 35, 37
Light 7, 86
Liver 20, 27, 62, 64, 67, 68, 81, 82
Loxus 14
*Lucidator* 3, 7, 8n47
Lust 121, 125, 131

Macrobius 22n30
Macrocosm 26
Magi 103, 106
Magic 4, 93, 101
Males 14, 71, 131
Mania 59
*Materia medica* 10
Materialism or materialist 1, 3, 5n30, 46, 73, 84, 90, 92, 103
Mathematics 33
Maurus of Salerno 22
McVaugh, Michael 33
Mechanical arts 17, 29, 78, 80, 94
Melancholy or melancholic 57–60, 97
Memory 19, 20–23, 35, 62, 75, 78–80
Mendicant friars 30
*Metaphysica* (al-Gazel) 108n79
*Metaphysica* (Aristotle) 32n99, 112
Metaphysics (philosophy) 13, 31, 34, 112
*Meteorologica* (Aristotle) 5n33, 24, 36, 47, 48n15, 48n16, 48n18
Michael Scot 35n122, 40
Microcosm 26
Mind 56, 73, 97, 106
Miracles 9, 28, 30, 35, 43, 108
Mixtures 23, 24, 45, 69
Moisture 65, 117
Monks 30
Monte Cassino 18
Montpellier 31, 36–38
Morality, Christian 42
Motive powers 77, 78, 82
Muscles 79, 82

Nardi, Bruno 72, 84, 87, 90
Naturals 10, 19, 20, 23, 27, 47, 62, 112
Necessity 14, 15, 82, 122
Nemesius 25, 26
– *De natura hominis* or *Premnon physicon* 21
Nerves 19, 22, 27, 55, 63, 70, 79, 82
Neutral state 23
Non-naturals 10, 20, 23, 112, 131
Normandy 25
Nutrition 19, 61n3, 66, 70, 77, 81, 121
– Nourishment 68
– nutritive powers or soul 69, 72, 77, 78

Occult properties, forces, or causes 4, 7–8, 93, 100, 101
– occult senses (*see* senses, internal)
Operations (one of the naturals) 10, 19
*Organon* (Aristotle's logical texts) 34
Organs 45, 54–56, 66–68, 70, 71, 74n69, 75, 79, 82, 91, 118
– formation of organs 64
– heterogeneous organs 48
– principal organs (*see* parts or organs, principal)
Oxford 6

Padua vii, 2–4, 11, 15n84, 31
Pain 13n69–70, 103, 118
*Pantegni* (*see* Haly Abbas)
Paris 3–4, 6–9, 11, 26, 28n72, 31, 33–37, 38, 41–43, 96, 113
Parts or organs, principal 40, 61, 63, 66, 67, 81 (*see also* organs)
*Parva naturalia* (Aristotle) 5n33, 34, 35n121, 80
Passions 14, 20, 25, 91, 97, 98, 113, 117, 121, 124–126, 128, 129 (*see also* affections and emotions)
Patient 93–95, 97, 102, 106, 109
Patristic sources 28
Perception 19, 20, 23, 75, 96 (*see also* apprehension, sensation, and senses)
Perfection, human 8, 43–44, 55, 120, 122, 132

- perfection of moral virtues 114
Peter Comestor 28–29
Peter of Spain 33, 36, 40
Petrarch, Francesco, *Invective contra medicum* 1, 94
Pharmaceuticals 4
Philaretus, *Pulses* 18n10
Phlegm or phlegmatic 1, 58, 97
*Physica* (Aristotle) 22n30, 24, 107n77
*Physica* (philosophy of nature) 23–25, 27
*Physiognomy* (as a science) 14–15, 87
*Physiognomy* (Ps.-Aristotle) 14–15, 88
Physiology 1, 6, 17, 19–22, 26, 27, 32, 40, 52n32, 57, 75, 89, 11, 114, 115, 119, 126, 128
Pity 49
Planets 7, 103, 131 (*see also* astrology and conjunctions)
- appeals to planets 102–103, 106
Plants 21
Plato 86
- Platonic cosmology 25–27
- Platonic views on soul 19, 21, 66
- *Timaeus* 26
Plurality of forms 72
Poetry 58–59
Poisons 4
Polemic or polemical 8, 43, 76, 122
Polemon 14–15
Politics 58, 78
Pope 3
- Gregory IX 35
- Gregory the Great 28–29
- Innocent III 30
Potency of matter 3, 50, 83, 92
Practice, medical 10, 17, 19, 20, 29, 37, 41, 45, 94, 100
*Praeter* naturals 10
*Premnon physicon* (see Nemesius)
Priests 29n81, 30
Principal parts (*see* parts)
*Problemata* (Ps.-Aristotle) 5, 6, 11–13, 24, 52, 57, 58, 87n53, 89n61, 91, 111–115, 118n35, 129
Prognostication 95
*Prognostics*, Hippocratic 18n10, 33, 94, 95
- Galen on the Hippocratic *Prognostics* 33, 38, 95
Prophecy 35, 108
Prudence 59, 78, 80, 113, 122
Pulse 115–116
*Pulses* (*see* Philaretus)
*Questiones naturales* (*see* Adelard of Bath)
Qustā ibn Lūqā, *De differentia animae et spiritus* 27, 62
- *De physicis ligaturis* 100

Rashness 23
Regimen 29, 33, 46, 60, 119, 120n44, 122n56, 129, 131
Relics 28
Reminiscence 78, 80
Renaissance, Twelfth Century 18
Renan, Ernst 84
Reproduction 35, 61
Respiration 35, 61
Rest 20
*Rete mirabile* 62
Rhetoric effecting health 93, 94, 106
Robert Grosseteste 25
Robert, Aurélien 84, 109
Roger Bacon 3, 7, 25, 32, 36, 46, 79n13, 97, 101

Sadness 13n69–70, 82, 97
Salerno 18, 21, 24n44, 25, 31, 34
Sanguine complexion 58, 60
Scripture 35
Scrovegni Chapel vii, 15
*Secretum secretorum* (Ps-Aristotle) 5n33, 6, 46n10
Semen (see also sperm) 68, 70, 115
Seneca (pseudo) 123, 127n77
Sensation 21, 40, 61, 79, 82 (*see also* apprehension and perception)
Senses, external or manifest 79, 80, 97, 120
Senses, internal, interior, or occult 8, 78, 79, 80, 88, 91, 96, 97, 106
Sex 29, 121, 131
- sexuality 8
- sexual desire 9, 43
Sickness (*see* disease)
Sienna 40
Signs 10, 14, 15, 19, 23, 95, 114, 115
Sin 28, 29
- Lucifer's sin 128
- original 26, 30
Siraisi, Nancy 2n6
Skin 14
Sleep (and waking) 20, 35, 52, 103
Slowness 23
Socrates 32, 98, 102

Species, intelligible or intellectual 85, 90, 95n20, 97–99 (*see also* intentions)
- sensible species 97
Sperm 67–69, 74, 82n25
Strength 52
Subalternation 32
Supernatural 8, 26, 105
- supernatural complexion 57

Taddeo Alderotti 34n118, 37–41, 55n47
*Tegni (Ars Parva)* (Galen) 10, 17, 18n10, 23, 24, 33, 38, 39, 46n9, 117n29
Temper, bad 23
Temperance 96, 113, 119 (*see also* balance)
- temperate complexions 59, 91, 92, 123
- temperate heat 65
- temperate parts 45
- temperate people 120, 131
Tendencies 14, 30, 49, 52
Testicles 81–82, 115
Theology 8, 31, 33n110, 36, 43, 46
- faculty of 3, 35, 36, 113
- theologians 6, 8, 28
Theophilus, *Urines* 18n10
Theory, medical 1, 2, 10, 17–29, 3, 34, 37, 40
Therapy 28, 30
Thierry of Chartres 26
Thomas Aquinas 36, 53n37, 72–73, 77n6, 79n13, 84, 87, 88n56, 101, 119n39, 123n59
Thomas of Strasburg 3n14, 9n52
Thorndike, Lynn 109

*Timaeus* (see Plato)
Toledo 32
*Topica* (Aristotle) 24, 35n123
Transubstantiation 9, 43
Treatments 23, 99
Trembling 115–116, 119
Trinity 63
*Trivium* 34
Turisanus (Pietro Torrigiano) 39, 71n57

Urine 115
*Urines* (see Theophilus)
Urso of Salerno 23–24, 94n5, 100

van der lugt, Maaike 24
Veins 19, 27, 61, 82
Vescovini, Graziella 84
Vices vii, 59, 111, 113, 118, 119, 121, 126
Volition 19

Wallis, Faith 23
Will, free 21, 42, 114n14
- willed and unwilled actions 129 (*see also* involuntary)
William of Conches 26
William of Moerbeke 34n119, 35n121. 35n122, 90n64
Wine 115, 118
Wisdom or wise 46, 49, 59, 60, 105, 113, 122
- divine wisdom 3n12
Woman 48
Womb 67–70